Philosophy of Physics: A Very Short Introduction

VERY SHORT INTRODUCTIONS are for anyone wanting a stimulating and accessible way into a new subject. They are written by experts, and have been translated into more than 45 different languages.

The series began in 1995, and now covers a wide variety of topics in every discipline. The VSI library currently contains over 650 volumes—a Very Short Introduction to everything from Psychology and Philosophy of Science to American History and Relativity—and continues to grow in every subject area.

Very Short Introductions available now:

WEATHER Storm Dunlop
THE WELFARE STATE David Garland
WILLIAM SHAKESPEARE
    Stanley Wells
WITCHCRAFT Malcolm Gaskill
WITTGENSTEIN A. C. Grayling
WORK Stephen Fineman

WORLD MUSIC Philip Bohlman
THE WORLD TRADE
    ORGANIZATION Amrita Narlikar
WORLD WAR II Gerhard L. Weinberg
WRITING AND SCRIPT
    Andrew Robinson
ZIONISM Michael Stanislawski

Available soon:

AMPHIBIANS T. S. Kemp
BIOCHEMISTRY Mark Lorch
CREATIVITY Vlad Glăveanu

SILENT FILM
    Donna Kornhaber
CITY PLANNING Carl Abbott

For more information visit our website

www.oup.com/vsi/

David Wallace

PHILOSOPHY OF PHYSICS

A Very Short Introduction

OXFORD

UNIVERSITY PRESS

Great Clarendon Street, Oxford, OX2 6DP,
United Kingdom

Oxford University Press is a department of the University of Oxford.
It furthers the University's objective of excellence in research, scholarship,
and education by publishing worldwide. Oxford is a registered trade mark of
Oxford University Press in the UK and in certain other countries

© David Wallace 2021

The moral rights of the author have been asserted

First edition published in 2021

All rights reserved. No part of this publication may be reproduced, stored in
a retrieval system, or transmitted, in any form or by any means, without the
prior permission in writing of Oxford University Press, or as expressly permitted
by law, by licence or under terms agreed with the appropriate reprographics
rights organization. Enquiries concerning reproduction outside the scope of the
above should be sent to the Rights Department, Oxford University Press, at the
address above

You must not circulate this work in any other form
and you must impose this same condition on any acquirer

Published in the United States of America by Oxford University Press
198 Madison Avenue, New York, NY 10016, United States of America

British Library Cataloguing in Publication Data
Data available

Library of Congress Control Number: 2020947757
ISBN 978-0-19-881432-0

Printed and bound by
CPI Group (UK) Ltd, Croydon, CR0 4YY

Links to third party websites are provided by Oxford in good faith and
for information only. Oxford disclaims any responsibility for the materials
contained in any third party website referenced in this work.

To Maia

Contents

Acknowledgements

This book draws extensively on a decade of teaching philosophy of physics to students in Oxford. I am especially grateful to Harvey Brown, whose ideas about spacetime had a substantial influence on Chapter 2, but insights from my wonderful students and wonderful colleagues have shaped every part of the book. I am also grateful to Latha Menon at Oxford University Press, to an anonymous referee, and to Hannah Wallace, for their careful and considered comments on the manuscript.

List of illustrations

Introduction

Until a few hundred years ago, 'the philosophy of physics' would have seemed a tautology. Physics was natural philosophy; the task of the natural philosopher was to understand the natural world. Aristotle's writings are maths and science as much as they are ethics and aesthetics; Newton described his work as a philosophy and contrasted it to the philosophies of his rivals. It is only in comparatively recent times that physics began to establish itself as a self-contained discipline and to separate away from philosophy writ large.

It was not the first discipline to do so, and it was not the last. It is often said that philosophy makes no progress, but to a large extent the creation of autonomous disciplines *is* how philosophy progresses. Mathematics in antiquity; physics in the Renaissance; biology after Darwin; logic in the early 20th century; computer science in mid-20th century; cognitive science still more recently; in each case, so much progress was made, so many controversies resolved, so many confusions clarified, that a self-contained subject was created and equipped to progress further. The philosopher Daniel Dennett defines philosophy as what we do when we don't know what questions to ask; when we understand enough to work out what the questions are and can start answering them, a new science buds off from philosophy.

Where this occurred recently—in cognitive science, say, or in logic—science and philosophy remain intertwined as disciplines. But physics has had three centuries to establish its independence, and now its institutional separation from philosophy is nearly complete: few professionally trained physicists learn much philosophy; few philosophers know more than the rudiments of physics. How, then, can there remain a philosophy *of* physics?

The clearest and simplest reason is that while the conceptual foundations of physics are clearer by far than they were before Newton's time, there remains much that we do not understand. Physics is not simply mindless calculation: good physicists are alive to the conceptual questions and paradoxes that arise in their work. Indeed, very little good calculation is mindless: there is no sharp divide between the concrete, predictive aspects of physics and its conceptual foundations. This is the first task for the philosopher of physics: much of philosophy of physics is continuous with this aspect of physics itself, and philosophical clarity can help move physics forward.

The second reason is that the methods of science, and the attitudes that scientists, philosophers, and people in general should take to them, matter a lot, and cannot be understood in ignorance of the details. We get at best limited insight into the scientific method, or the attitude we should take to scientific theories, if we consider them abstractly: one important task for the philosopher of physics is to look at the results and the processes of physics and see what lessons they teach us about science more generally. Again, this is not simply a task for dispassionate outside observers of physics: an understanding of the scientific method is important for good science, and is becoming more and more so in modern physics.

The third task for a philosopher of physics is again continuous with physics, but in a different sense: to get the best understanding we can of what the world is like, given the physical

2

theories we use to describe it. Traditionally, understanding the deep nature of the world was the task of *metaphysics*, but in modern times that understanding relies critically on our best physics theories—yet those theories do not wear their meaning on their sleeve. In this sense, philosophy of physics provides a bridge between the metaphysician and the physicist—or, put another way, philosophy of physics tells us how to do a metaphysics that is scientifically informed.

A common theme here is that philosophy of physics is interdisciplinary, sitting between physics proper, mainstream philosophy, and the general philosophy of science, and communicating ideas and insights between them. Some philosophers of physics work in philosophy departments and have 'philosopher of physics' or 'philosopher of science' as their job title, but many others are physicists. (Einstein, without question, is the leading philosopher of physics of the 20th century.) This is a book about the philosophy of physics as a subject, not as an institution within 21st-century academia.

Another theme is that the devil is in the details. Philosophy of physics is rarely concerned with physics as a whole but with particular areas within it. Given a field in physics we can consider the conceptual—that is, philosophical—questions that arise in that field, and the problems in each sub-field are distinctive, even if they are never as cleanly separated as it might appear on the surface. There are many examples, some of which have become important in very recent work and engage with the frontiers of current physics. For instance:

- The philosophy of *quantum field theory* addresses conceptual questions rising out of our most fundamental current theories, those which underpin the remarkable results obtained at the Large Hadron Collider (LHC) in Geneva. For instance: are the constants of nature finely tuned in some way to bring about the large complex Universe we see around us? Or is that a meaningless question even to ask? And what scientific method makes sense to develop our theories beyond those tested at the LHC, into a realm where theories seem all too plentiful and evidence seems scarce?

- The philosophy of *cosmology* explores how to do the science of the whole Universe. When there is only one Universe to observe, how can we understand, let alone test, theories that say that one Universe is more likely than another? Is it scientific to hypothesize that our Universe is just one member of a vast multiverse? And how do we select between different cosmological models given the complex and indirect evidence that bears on which is correct?

- The philosophy of *quantum gravity* is concerned with the 'final frontier' in modern physics: the quest for a *rapprochement* between general relativity, our best theory of gravity, and quantum theory, the framework in which all our other physical theories are written. Is string theory—the leading contender for that *rapprochement*—good science, bad science, or not science at all? What is at stake in the astonishing claims that black holes—from which, naively, we expect nothing to escape—behave in important ways like ordinary hot bodies, and how can we evaluate those claims given the remoteness of any prospect of experimental test? And how are the conceptual paradoxes about quantum black holes to be resolved or dissolved?

But the most important examples, at least for most of the foundational and philosophical work of the last fifty years, are somewhat older and more general theories. The bulk of work in philosophy of physics is concerned with three areas: the philosophy of spacetime (which provides the concepts by which space, time, and motion, perhaps the central concepts in physics, can be understood); the philosophy of statistical mechanics (which underlies the relations between physical theories on different scales, and stands between our most fundamental physics and almost any test or application of that physics); perhaps most famously, the philosophy of quantum mechanics (which tries to understand, and perhaps to change, the astonishingly strange language in which is written most of the last century of physics, from particle accelerators through transistors to the earliest moments of the Universe).

This book is structured around these areas. Chapter 1 begins with some fairly general questions about the scientific method and the nature of scientific theories more generally. Chapter 2 addresses the question 'what is motion', in a historical context: the creation of mechanics by Newton and others in the 17th century. (It is a common theme in philosophy of physics that we need to pay attention to the history of ideas, and to how great physicists were led to the strange concepts that they developed.) Chapter 3 brings these considerations forward to the theory of relativity, to the supposed 'paradoxes' that this theory contains, and to the deeper understanding of space, time, and motion that it provides. Chapter 4 considers how statistical mechanics helps us understand the relations between theories on large and small scales, and how that understanding seems to involve new concepts—irreversibility, probability—which are absent from, and even seem inconsistent with, our more fundamental theories. Chapters 5 and 6 address quantum mechanics—first presenting the paradoxes it raises, then considering how they might be resolved, and finally asking why this matters to physics more generally.

A third theme in philosophy of physics is that maths matters. Theories in modern physics are formulated in fairly abstract mathematics, and it is not possible to completely understand—let alone contribute to solving—the philosophical problems of physics without understanding those theories in full detail, maths and all. (It's not surprising, given this, that philosophers of physics usually—and increasingly—have graduate training in physics, even when they work in philosophy departments.) In a book like this, it's not realistic to give a full account of any of modern spacetime theory, statistical mechanics, or quantum theory—let

5

alone all three—and so I can't pretend that you will gain a complete understanding of the philosophical issues from this book alone. But I hope there is enough detail and depth to help readers trained in philosophy to understand how issues from physics affect deep questions relevant to them, readers trained in physics to connect the conceptual issues I discuss to their technical training, and all readers to get some insight into the central problems of the field and why they are important.

Even in the three core areas I discuss, and even setting aside many topics too technical to include in a book like this, there are many more interesting questions than I have space to consider. Inevitably, my own judgements—both about the most important questions, and about the most promising attempts to answer those questions—played a major role in the choice of what to discuss and how to present it, and while often I note that a question is 'contested' or 'controversial', and present multiple possible answers to that question, still I am sure that readers familiar with these topics will see some of their favourite objections elided, or their favourite positions skimmed or skipped. To them I can only apologize; for readers not already familiar with philosophy of physics—my target audience—I have included various further readings, offering more details and alternate perspectives.

Chapter 1
The methods and fruits of science

Physics is a part of science, but there is an important sense in which the *philosophy* of physics is not just a part of the philosophy of science. Traditionally, the philosopher of science has stood a little apart from science, studying the practice of science as an interested outsider: their concern has been with the scientific method in general, and with comparably general questions as to the reliability and value of the product of science. Philosophers of physics are usually concerned with much more specific questions: not about science in general, nor even about physics in general, but with conceptual questions raised by specific physical theories: what does the general theory of relativity tell us about space and time; how should we understand the second law of thermodynamics; is there something fundamentally wrong with quantum theory? And while they have not been reticent about criticizing physicists' methods in particular cases, they have not focused on more general questions of physics methodology.

For all that, the questions asked by general philosophy of science—and the tentative answers that it has provided—are a crucial backdrop for philosophy of physics. So my task in this chapter is to give an introduction to the scientific method, and to some puzzles about how to think about scientific theories in general.

The scientific method—from induction to falsification and beyond

Let go of an apple, and what happens? It falls. Do the same thing with another apple, or the same apple at a different height; it falls. The same for other objects: pears, bricks, cats, children. Experience gives us many, many instances of "If such-and-such object is unsupported, it falls" and precious few of "If such-and-such object is unsupported, it *does not* fall", and so, perhaps we are entitled to take "All unsupported objects fall" as a tentative conclusion from our observations.

Call this *enumerative induction*. It is probably the simplest and oldest model of empirical knowledge: collect lots of observations of the form 'this is an X that is also a Y', fail to observe 'this is an X that is not a Y', and so infer 'all Xs are Ys'.

Maybe this is a good model for how infants learn about gravity. (I really doubt it, frankly, but I'm no cognitive scientist.) *Maybe* it captures some of how a new field of science proceeds in its infancy. Even as a caricature of the method of physics, or any other mature science, it is hopeless.

Why? Partly because there is no such thing as 'just observing'. The world is rich and complicated and we can look at it in indefinitely many ways: we need to make some judgements as to *what* to observe. That's true even for the observations we make with our eyes and ears, but it's magnified a hundredfold in modern physics, where millions or billions of dollars are spent to develop the capacity to make very specific observations that are by no means chosen at random. And partly because the form of a scientific theory is far subtler and more complex than 'all Xs are Ys'. The actual form of Newton's theory of gravity is not, even schematically, 'All unsupported objects fall', but rather 'The gravitational force on one body due to another is proportional to the product of the masses divided by the square of the distance between them and acts along the line connecting the two'. That's not something you can mindlessly read off the world.

Really, these are two sides of the same coin: what enumerative induction gets wrong is that it mixes together the process of *coming up with* a theory (what philosophers call the 'context of discovery') with the process of *collecting and assessing evidence for* a theory (the 'context of justification').

A highly influential alternative was developed in the 20th century by the philosopher Karl Popper (one of the very few philosophers of science who most physicists have heard of). In the simplest and best-known version of Popper's approach, the scientific method is a two-step process:

1. Come up with a theory (never mind how);
2. Attempt to *falsify the theory*: that is, test some prediction the theory makes.

If the theory fails the test, it's falsified: throw it out, and go back to step (1). If it passes, keep testing it in different ways.

Call this approach *falsificationism*. Unlike enumerative induction, it *is* a caricature of the scientific method—and, like all caricatures, it captures some of the central features of its subject matter, but its details shouldn't be taken too literally, and if they are, it is likely to mislead.

We can see this through a real-world example from 19th-century physics. According to Newtonian gravity, planets orbit the Sun in ellipses; the *perihelion* of an orbit is the point of closest approach to the Sun. Also according to Newtonian gravity, in the absence of any other planets, a planet's perihelion is at the same point in space on each orbit—but since the other planets are *not* absent, in practice the perihelion is a bit further around the Sun on each successive orbit. That is: it *precesses*, by an amount that Newtonian gravity tells us how to calculate.

When physicists did that calculation, they found that two of the seven then-known planets—Mercury and Uranus—showed a discrepancy between the value the theory predicts and the value they measured. The measured values are tiny, and so are the discrepancies: for Mercury, whose closest approach to the Sun puts it 44 million kilometres away, the predicted precession is about 3,000 kilometres per orbit, and the measured precession is about 20 kilometres less. But both the calculations and measurements were accurate enough, even in the 19th century, to be confident that the discrepancy is real.

According to falsificationism, that should be the end for Newtonian gravity. It made a prediction; that prediction was false; time to move on to the next theory! But that isn't what happened, and it isn't what should have happened. For one thing, Newtonian gravity had been highly successful for hundreds of years, with a huge number of successful predictions and informative explanations to its name: simply discarding it and starting afresh, in the absence of any concrete ideas of how to do better, would have paralysed astronomy. Even more importantly, it wasn't strictly true that Newtonian gravity was falsified by the discrepancy. For Newtonian gravity—like any theory in physics—only makes predictions with the aid of what philosophers call *auxiliary hypotheses*: about which planets there are, where they are, how massive they are, what other moonlets and asteroids and dust-clouds might be around, what non-gravitational effects might come into play, and even about how our telescopes and timepieces function. The anomalous precession might be because of a failure in the theory of gravity—but it equally well might be because of another distant planet that we didn't know about. Indeed, we can turn this logic around: *assuming* that Newtonian gravity is correct, where would another planet have to be in order for it to resolve the anomaly? When mathematicians asked and answered that question for Uranus, and then astronomers looked at that point in the night sky, they found the planet Neptune exactly where it should be.

What of Mercury? The same trick was tried: if some unknown planet was still closer to the Sun, then it could explain away the anomaly. This new planet was dubbed 'Vulcan'; no-one could find it, but that was hardly conclusive given that any such planet would be so close to the Sun as to be near-invisible in its glare. But with hindsight, the explanation was completely different: Einstein's *general theory of relativity*, a rival theory of gravity, predicted precisely the observed discrepancy with no need for any additional planet.

So: two apparent episodes of falsification; with hindsight, one was a triumph of Newtonian gravity, falsifying not the theory but our auxiliary assumptions about the solar system and leading to the discovery of the eighth planet; the other was a true falsification, explained by the wholesale replacement of Newtonian gravity with a new and improved theory. But *only* with hindsight can these distinctions be drawn: there was nothing inherently unreasonable about the idea of Vulcan, and no improvement in scientific method could or should have told scientists not to postulate it.

There is no settled consensus on how to tell a positive story of the scientific method that improves on falsificationism; there is not even consensus that it can be done at all. (The philosopher of mind Jerry Fodor offers this tongue-in-cheek suggestion: 'Try not to say anything false; try to keep your wits about you.') But some common themes in many accounts can be found, and will suffice for our purposes (here I follow philosophers Imre Lakatos and Thomas Kuhn, glossing over many differences in their accounts):

1. We should not think in terms of static theories, conjectured once-and-for-all and only thereafter tested: but instead of ongoing research programmes (Lakatos) or 'paradigms' (Kuhn) in which a common core of theory is used to explain phenomena via a collection of auxiliary hypotheses which can be altered to account for successively collected evidence.

2. Research programmes progress precisely through the discovery of anomalies which are then explained. The gold standard for such explanations is that they lead to novel predictions that are then confirmed (like Neptune).

3. Over time, unexplained anomalies can build up, and/or the changes to the auxiliary hypotheses needed to explain the anomalies become increasingly contrived, ad hoc, and unsuccessful in giving rise to novel predictions. The research programme is degenerating (Lakatos); the paradigm is in crisis (Kuhn).

4. Even so, we seldom if ever abandon a research programme except when some more successful rival is available. Research programmes are tested not simply against the world, but also against other research programmes. (It was not until the success of general relativity—a new research programme—that Newtonian gravity was regarded as falsified.)

The demarcation problem: when is something scientific?

Popper's own interest in falsificationism, and in the scientific method, was only partly for its own sake. He also sought a *criterion* for when an approach to knowledge-collection counted as science, and found it in the requirement of falsifiability (so, supposedly, neither Freudian psychology nor Marxist economics— two of his bugbears—counted as scientific). Modern physicists often seem to do likewise: to dismiss a question, or sometimes an entire sub-discipline (like string theory) or field of study (like philosophy!), as 'unfalsifiable' and thus unscientific.

We have seen that taken literally this is too simplistic: strictly speaking, no theory is falsifiable in isolation. But there is something in the idea nonetheless: what matters seems to be not whether a theory is falsifiable per se but whether evidence bears on it. So a question (such as: which of these theories is correct?) is a scientific question if it is amenable to the methods of science, which ultimately rest on evidence.

We can see how this plays out through a very recent case-study: the debate over the last thirty-odd years about the existence of dark matter. In spiral galaxies like our own, the visible matter is mostly in the form of stars and interstellar gas and dust, and we can use Newton's law of gravity to work out how fast stars in the galaxy ought to be orbiting (the 'rotation curve' of the galaxy) given the distribution of that matter. It has been known since the early 1980s that there is a discrepancy between the predicted and measured values. (And replacing Newtonian gravity with Einstein's general relativity makes no difference here: the anomaly persists.) Similar anomalies showed up in larger-scale observations of entire clusters of galaxies.

The main proposed explanation of this anomaly was so-called 'dark matter': matter not visible to our telescopes and detectable only through its gravitational influence. In itself, there is nothing odd about the idea of dark matter: stars are visible because they blaze with light, and matter that isn't self-illuminating can be hard to see. (The other planets of our solar system would count as 'dark matter' to an observer in a distant solar system—the Earth would have, too, until we started transmitting radio and TV signals.) But to explain the rotation curves, the dark matter needs to outweigh the stars, gas, and dust by a large fraction, far too large for such mundane explanations. To this day, we know almost nothing about what dark matter might be, and direct searches for it have failed.

And for that reason, a minority of physicists have explored the idea that there is no dark matter after all, and that instead there is something wrong with gravity. Their rival theory, MOND (for 'modification of Newtonian dynamics') had considerable initial success in explaining the rotation curves, and at least some success in explaining other apparent evidence for dark matter. In the terms we have discussed, dark matter is an auxiliary hypothesis within the existing gravitational research programme; MOND is a rival research programme.

This debate is now at least thirty years old, and it lacks, and for the foreseeable future will continue to lack, a decisive resolution, because no observation or experiment can plausibly *falsify* either approach. But this is not to say that it just a matter of taste which theory to prefer. The explanations offered by MOND are in some cases simpler than for dark matter, and account for the phenomena with fewer moving parts; in others, they are more complex, even contrived. And the discussion has not been static: new observations have required new refinements of the dark-matter models and of MOND's proposed new laws. My own assessment is that twenty years ago MOND was a highly plausible rival, but by now the level of contrivance and ad hoc modifications required to fit the data makes it most unlikely to be correct. And that assessment is shared by most of the astrophysics community, as can be seen by the sharply decreased interest in MOND now as compared to twenty years ago. Yet it is not conclusive: reasonable people can disagree here, and serious scientists continue to work on MOND. There may come a time when it *is* conclusive; then, and only then, would it be right to call support for MOND unscientific.

I should stress that dark matter/MOND is a fairly extreme example: the difficulty of doing experiments and making observations is much harder in these corners of astrophysics and cosmology than in most areas of physics, where the accumulation of experimental evidence is usually faster and more conclusive.

But even in that extreme case, we can see that the resolution of the disagreement proceeds through new developments of the theories and new proposals for observations: that is, it proceeds through recognizably scientific means.

Underdetermination, instrumentalism, and realism

If the choice between two theories is scientific, evidence must bear on it; but what about where two distinct theories make exactly the same predictions? Philosophers call this case 'underdetermination of theory by evidence': the 'underdetermination problem' is the problem of choosing between two such theories.

In a sense, the dark matter/MOND question is a case of underdetermination: no single experiment or observation tells us which theory is correct, because the details of the theory and the auxiliary hypotheses can be filled in in different ways to account for the same data. The apparent solution to the problem is simply to do more science: over time, the weight of evidence—we can hope—leads to an increasingly clear conclusion, even if there can remain residual philosophical questions as to just why that 'clear conclusion' can be trusted given that neither theory is strictly ruled out. (Philosophers call cases like this *weak underdetermination*.)

The more disturbing case would be where two theories make the same predictions not just with respect to the current set of evidence, but with respect to all possible observations and measurements. In this case—called *strong underdetermination* by philosophers—the methods of science do not seem to get traction.

Once, many philosophers of science thought that this was logically impossible. According to *positivism* (or *operationalism*, or *instrumentalism*—once again, I set aside subtle differences between these positions), to understand the content of a scientific theory we need to distinguish its *observational* claims from its *theoretical* claims. And here, 'observational' is used in a pretty strong sense, to mean something like 'expressible in the language of everyday objects' or 'testable with unaided human senses'. 'This detector will display the number 5.228', for instance, is an observational claim. To the positivists, *non*-observational claims cannot be understood independently of their observational consequences: to say, for example, 'atoms are made up of electrons and protons' is just to say 'if I make *this* measurement I'll get *this* result, if I make *that* measurement I'll get *that* result,…'. In effect, the content of a scientific theory just is the collection of observational claims it makes: the rest of the theory is just a calculational tool to get from one set of observations to another.

The appeal of positivism to philosophers in the early 20th century—and to some physicists even today—is that it dismisses as meaningless any claim about a theory that does not have experimental consequences. In particular, it rules out strong underdetermination by fiat: if two theories make the same observational claims, they're the same theory—just presented differently. Positivism lets us cut through a lot of verbal confusion and get to the actual, scientific questions.

What's not to like? For one thing, it's a poor fit to the way working scientists describe their projects, to themselves and others. Astronomers usually describe themselves as building radio telescopes in order to learn about stars and galaxies, not because radio telescopes are intrinsically interesting to them. When particle physicists exhort government ministries to spend large sums on building particle accelerators, their case rests on using the particle accelerator to discover deep truths about the Universe; that case would be pretty severely undermined if the 'deep truths about the Universe' needed to be understood just as indirect claims about the workings of the particle accelerator itself.

But there is a deeper reason: the distinction between observational and theoretical claims is much harder to make than it looks. We have already seen part of the reason: because of auxiliary hypotheses, no scientific claim in isolation has observational consequences, and so my schematic list of observational consequences of 'atoms are made up of electrons and protons' is a fiction. Ultimately, the observational claims of a scientific theory rest on its theoretical claims, and cannot be fully understood without them.

Actually, it's even worse than that: the distinction between 'observational' and 'theoretical' claims can't really be made at all. In philosophers' jargon, observations are *theory-laden*: even to describe an observation, we need the language of theory. 'This detector will display the number 5.228'...*which* detector? Only a detector built the right way—and 'the right way' inevitably involves theoretical ideas. Terms like 'laser' or 'radio telescope' or 'particle accelerator' just can't be understood without understanding the theory in which those terms are used (a 'laser', for instance, is a beam of coherent light generated by stimulated emission—but now let's talk about what 'coherent' and 'light' and 'stimulated emission' mean...and so on).

That might seem to put paid to the threat of strong underdetermination: if we can't separate out the observable and unobservable parts of a theory, we can't even make sense of the idea of distinct theories with the same observational consequences. At worst, we might have a case of weak underdetermination, like the dark matter/MOND case, where evidence does bear on the debate but at present it does so inconclusively. Nonetheless, there are still ways in which strong underdetermination might occur.

The first is boring: we can cheat. If theory X, say, involves electrons, theory X* might be, 'All observations occur as if electrons exist, just as X says, but actually they don't exist'. It's actually quite subtle to say just what's wrong with a theory like X* (it's an example of what philosophers call *scepticism*), but the problem isn't really specific to science. (Consider: all observations occur as if the Eiffel Tower exists, but actually it doesn't exist.) If we're interested in the philosophy of *science*, and not just in general philosophical puzzles, we can set this case aside.

The second occurs when two theories are identical *except* for an extra bit in one theory that doesn't do any explanatory work. If theory X is some successful scientific theory, X* might be obtained by adding to X a new particle—the 'irrelevanton'—that doesn't interact with any of the other particles. In that case, the irrelevanton just complicates X without adding any scientific power to X, and so it seems pretty reasonable scientifically to stick with X. (Again, exactly *why* this is reasonable can be a bit subtle. A hundred years ago, philosophers would have said that the irrelevanton is unintelligible. Today, they would be more likely to just say that we have no reason to believe it exists.)

The most interesting case brings out a sharp difference between the methods of physicists and (most) philosophers. To a working physicist, a theory is ultimately expressed in mathematics: a collection of mathematical structures describing apparent possibilities, some equations that say which apparent possibilities are really physically possible. But the world does not just seem to be mathematics, and there is at least a strong temptation to require more of a theory: some account of what there is in the world according to the theory, how it behaves, what causes what, what the explanations are. If so, that raises the possibility that two distinct theories might have the same mathematical structure (and so, given the way physics is done, the same observational consequences).

Is this a real possibility? We have seen that instrumentalists say no: to them, any two theories with the same observational consequences are really the same theory described two different ways. (But we have seen that instrumentalism is not really viable.) *Standard scientific realists* say yes: according to them, two distinct theories can share the same mathematical structure. *Structural scientific realists* (or just *structuralists*) also say no: they hold that two theories with the same mathematical structure are really different descriptions of the same theory. The structuralist approach is closer to the tacit assumptions used in physics, and I mostly adopt it here, but there are many unresolved questions about how it is to be understood and differentiated from the standard approach.

Scientific realism and theory change

What does the term 'scientific realism' in structural or standard scientific realism refer to? To the view, standard among most philosophers and (at least tacitly) most scientists, that the success of our current scientific theories gives us good reason to think that they are correct (and not merely useful gadgets to make predictions). Electrons, or quarks, or black holes, cannot be directly observed—that is, you can't see, hear, or touch them—but (say scientific realists) we still have good reason to think that there are such things.

There are two main (and related) arguments for scientific realism. Both start with the clear evidence that our best theories of physics are really, really effective at describing the physical world. (No-one worth taking seriously is an astrological realist, because astrology just isn't a successful theory.) The first argument is then that there is no remotely plausible way to understand why those theories are so successful, other than by assuming that they are at least roughly correct. This is sometimes called the *no miracles argument*, following philosopher Hilary Putnam's observation that it would be a miracle for scientific theories to work so well if they weren't true. For instance, Newtonian gravity predicted the presence of Neptune, in a specific spot in the sky; we looked; there it was.

That's unsurprising *if* Newtonian gravity were correct; if it weren't, it would be a miraculous coincidence.

The second argument for scientific realism is that since (as we have seen) there is no real way to separate observational from non-observational content in a theory, there is no principled way to accept a scientific theory as observationally adequate without just accepting the whole theory. So when we are led through good scientific methodology to formulate, test, and accept a theory, we're already committed in so doing to accepting the theory *as true*.

Scientific realism might seem just obvious: isn't doubting our well-established theories a sort of anti-scientific scepticism? But there are reasons to treat it cautiously. The first we have already met: the threat of underdetermination. If we have two theories that make the same observational predictions but which contradict one another, then (at least) one of them must be false—in which case the no-miracles argument can't be right. (This is one reason philosophers have been so interested in whether strong underdetermination actually happens.)

The other reason comes from the history of science—especially physics. Repeatedly in that history, an established theory has been overthrown, even though it was highly successful at making predictions. Newtonian gravity, for instance, is in an important sense *wrong*, replaced by the general theory of relativity; so much, then, for the idea that the detection of Neptune would be miraculous if the theory was wrong! In general, science seems to proceed at least partially through revolutionary steps in which old theories are replaced by new ones that contradict central claims of the old theory. For instance, physicists used to think heat was a fluid; now they think it's random motion of molecules; they used to think light was a vibration in an all-encompassing 'aether'; now they think it can exist in the absence of any such thing. This *pessimistic argument* has historically been the main objection raised to realism.

But the extent of theory change can be overstated, even in physics. Newtonian gravity is still taught to students, even now, and not just as a warm-up exercise (recall that MOND is a 'modification of *Newtonian* dynamics'). The standard reason given is that Newtonian gravity is a very good approximation to general relativity in certain circumstances. The question for scientific realists is then: does that just mean that the predictions of one theory are good approximations for the predictions of the other? Or is the content of the old theory still approximately correct in the new theory?

The distinction between standard and structural realism is important here too. When physicists say that one theory approximates another, they normally mean that the mathematical structure of the first theory remains approximately realized by the second theory. So if the structuralist is correct that a theory is completely given by its mathematical structure, then it is not too difficult to see how the old theory could still be approximately correct even though the new theory is better. (The equations for heat flow don't care whether what is flowing is a fluid or a quantity of vibration; the equations for light have roughly the same structure in the aether theory as in modern electromagnetic-wave theory.) The standard realist has a more difficult challenge.

That concludes our brief tour of general philosophy of science. The key points, which will come up repeatedly in what follows, are:

- Falsification is a big improvement on induction as a description of the scientific method, but it is still only a crude approximation—a given observation usually only falsifies a theory given a host of background assumptions. So there is no simple, one-off test for when something is science: we have to look at how a scientific research programme progresses or regresses.

- Underdetermination—where two different theories give the same predictions—is rarely an all-or-nothing affair, because the distinction between theoretical and observational claims is blurry. Apparent cases of underdetermination often get resolved over time as one theory turns out to be more powerful as a framework. The only realistic cases of exact underdetermination seem to be where two theories are mathematically equivalent; in these cases, physicists—and some, but not all, philosophers of physics—regard them as the same theory.

- Instrumentalism—the view that a scientific theory is no more than its empirical predictions—is widely rejected in philosophy of science, again because it relies on a sharp distinction between empirical and theoretical parts of a theory which scientific practice does not sustain. The predominant 'scientific realist' position in science and philosophy takes scientific theories literally as attempted descriptions of what is really going on in a system, even when some parts of that description are invisible to the naked eye.

In Chapter 2, I will begin discussing philosophy of physics proper—beginning with the philosophy of space, time, and motion.

Chapter 2
Motion and inertia

Cast your mind back to the 17th century—the birthplace of modern physics, midwifed by René Descartes, by Gottfried Liebniz, by Galileo Galilei, above all by Isaac Newton. This was a time when no distinction really existed between philosophy and science—when 'natural philosophy' just *meant* 'science'—but even so, it might be surprising to modern readers to realize how fixated thinkers of the time were with a quintessentially philosophical question: *what does it mean for something to move?* In this chapter we'll see how physics cannot be done without some kind of answer to this question, and how trying to answer it teaches deep truths about the nature of space and time. And in Chapter 3, we'll see that those truths are altered by the physics of the 20th century—but not transformed out of recognition, so that many of the deepest principles of today's physics still rest on insights that can be learned from the physics of the 17th century.

Rest and motion

Here's a basic posit of Newton's physics, learned by every schoolchild:

> *Newton's First Law* (naïve form): An object on which no force acts that is at rest will remain at rest; or, if moving, will remain in motion at a constant speed and in a straight line.

But what do 'rest' and 'motion' actually mean here? The most straightforward answer is *relative motion*: particle A is at rest *relative to particle B* if the distance between A and B is unchanging. We can make this more detailed, and more useful: given some extended object—say, the Earth—motions relative to that object give a pretty detailed description of a particle's movement. We can say of a particle, for instance, that it's 3,100 metres above the Earth's surface, at a latitude of 34.0522° N and a longitude of 118.2437° W, and if we know how fast those numbers are changing, we know how fast the particle is moving, relative to Earth, in the north/south, west/east and up/down directions. In this way, the Earth defines a *material reference frame* that can be used as a standard of motion.

But relative motion—even motion relative to a material reference frame—can't be what 'motion' means in Newton's First Law. Body A might be moving with respect to body B, but stationary with respect to body C, so that it's not uniquely determined whether a body is at relative rest or in relative motion. But the First Law just talks about *motion*, not *motion relative to this body or that*. (Newton himself called this *absolute motion*, or *true motion*). Similarly, Newton's Second Law says that the acceleration of a body is proportional to the force acting upon it, and acceleration—which is the rate of change of velocity—is again an absolute, with no mention of which other body it's relative to.

(A quick reminder: *velocity* is speed plus direction. If you're driving along at 30 miles per hour and you turn left without slowing down, your velocity has changed even though your speed hasn't.)

We can see the problem another way, by asking whether Newton's laws are true of motions measured in the material reference frame of the Earth. That's an empirical question, with an empirical answer: no, not exactly. They fail to hold in some circumstances because the Earth itself is not at rest: it is rotating, sometimes it experiences earthquakes, and so forth. But even to speak of the Earth as moving, we can't use the Earth itself as the body relative to which motion is defined.

So what can we use? Astronomers often used to refer to the *fixed stars*—the heavens wheel across the night sky as time passes, but really it is the Earth that is rotating, against the fixed background of the stars. And indeed, using the stars as a material reference frame works much better than using the Earth. But it is still imperfect, for a similar reason: the stars are not, after all fixed (and Newton and his contemporaries knew as much). They move among themselves, as they collectively revolve around the centre of the Galaxy, and as the Galaxy itself falls towards its neighbour.

There is no simple solution here. If we want a material reference frame sufficient to define the motions that satisfy Newton's laws, it must be defined with respect to bodies that do not in any way move among themselves—a *rest frame*, if you like. And there are no such bodies.

Newton himself believed—and forcefully argued—that the only way to define 'motion' adequately was to admit something else to our picture of the world, something additional to all of the moving matter, something which would persist even if the matter was to vanish: *absolute space*. Newton's own concept of absolute space was heavily theological—he referred to it as 'the sensorium of God'—but the scientific case for absolute space can be made with no theology at all: it is the thing that defines the rest frame, the unchanging reference standard against which motions can be defined.

This idea of space as a thing separate from matter is called *substantivalism* by philosophers, and contrasts with *relationism*, the view that all there is in the world is matter. Substantivalists believe space is a substance, a thing itself over and above the material contents of the world; relationists believe 'space' is just a pretty way of talking about the relations that hold between bodies. This might seem to be an arcane, even a semantic debate, but Newton's arguments show its significance for physics: if all that exists is matter, we don't seem to have any way to define the rest frame that we need to do physics.

Now, philosophers disagree with one another (shock!) as to whether absolute space is necessary for the idea of a background rest frame to make sense; for our purposes, though, what really matters is not the nature of that rest frame but the fact that physics seems to need that frame and that it provides a *background* against which physics plays out. It is not itself a player in the dynamical games of physics—it just sits there—but it provides the standards of measurement by which those dynamical games can be defined, for the material bodies that obey the laws of physics in the foreground. As a purely philosophical matter, physics didn't have to be like this—one can imagine a physics expressed entirely in terms of relative distances and their changes, and indeed models of such physics have been constructed—but to correctly describe the actual physics of the systems Newton wished to study, the background seems necessary.

The relativity principle

But there is something a little strange about using motion *relative to an absolute rest frame* as the basis of mechanics, rather than motion relative to other material bodies. After all, we can see material bodies, and so we can see whether something is moving relative to them. We can't see the rest frame. (In Newton's terms: the points of absolute space are invisible.) Using an unmoving, immaterial reference frame to *define* motion seems of little use as long as we cannot *detect* motion.

Newton himself was well aware of the problem: in his *Principia Mathematica* he writes

It is indeed a matter of great difficulty to discover, and effectually to distinguish, the true motions of particular bodies from the apparent [i.e., the relative motions]; because the parts of that immovable space, in which those motions are performed, do by no means come under the observation of our senses. Yet the thing is not altogether desperate…

He goes on to consider an ingenious thought experiment: a pair of heavy globes connected by a string. The relative motions don't tell us if the whole setup is at rest, or if it is whirling around at high speed: either way, the distance between the globes doesn't change over time. But—according to Newton's theory itself—the tension in the string will be higher if the globes are rotating around one another. Indeed, if the string is under tension, we can see whether pushing on the globes in various ways increases or decreases that tension, and so determine not just whether and how rapidly the globes are spinning, but about what axis.

This illustrates one of the general themes of Chapter 1: scientific observations are theory-laden, so that what can or cannot be observed and measured in a theory is not a simple matter to determine but depends on the details of the theory itself. In Newtonian mechanics, absolute rotation is observable, even though it isn't directly present to the senses, via its dynamical relation to things that are more straightforwardly measured.

But can all motions be so detected? No. Consider this famous passage from Galileo:

Shut yourself up with some friend in the main cabin below decks on some large ship, and have with you there some flies, butterflies, and other small flying animals. Have a large bowl of water with some fish in it; hang up a bottle that empties drop by drop into a wide vessel beneath it. With the ship standing still, observe carefully how the little animals fly with equal speed to all sides of the cabin. The fish swim indifferently in all directions; the drops fall into the vessel beneath; and, in throwing something to your friend, you need throw it no more strongly in one direction than another, the distances being equal; jumping with your feet together, you pass equal spaces in every direction. When you have observed all these things carefully (though doubtless when the ship is standing still everything must happen in this way), have the ship proceed with any speed you like, so long as the motion is uniform and not fluctuating this way and that. You will discover not the least change in all the effects named, nor could you tell from any of them whether the ship was moving or standing still

Galileo's point is that the absolute velocity of a system of bodies is not detectable by any means available to a scientist who is part of that very system, because the relative motions of the bodies are unaffected by their overall velocity. Only by relating the bodies to some external system can the motion be detected (hence Galileo's injunction to be below decks, where the moving sea cannot be seen).

Is this correct? It certainly sounds intuitive (especially in the modern era of air travel: the author has more than once forgotten whether the plane he is on has taken off or not, a difference in absolute velocity of some 300 metres per second)—but couldn't there be some subtle effect that sufficiently precise measurements could detect?

There could not, *according to Newtonian mechanics itself*. The fact that relative motions among a group of bodies are unaffected if they are all given the same velocity is something that can be derived from the equations of mechanics. In the parlance of physics, it is a *dynamical symmetry*, a transformation of a system that leaves unchanged the physics governing that system and so cannot be detected by physical processes within the system. Whether a given transformation is a symmetry of a system depends which theory correctly describes that system, and so ultimately is a matter of experiment—but for a *given* theory it is just a mathematical fact what its symmetries are. In the case of Newtonian physics, velocity boosts—transformations where all bodies are increased in speed by the same amount, in the same direction—are, provably, symmetries. And so Galileo is right: velocity boosts are undetectable in Newtonian mechanics.

The postulate that velocity boosts are symmetries is called the *principle of relativity*. In popular culture it is of course associated with Albert Einstein, but the basic idea is hundreds of years older. And it creates a potentially severe problem for Newton's physics: it tells us that, contra Newton's suggestion, it is in principle impossible, according to physics itself, to detect whether or not something is moving with respect to the rest frame.

Inertial reference frames

To review: we *seem* to have established that:

1. We need to posit absolute space to provide a 'rest frame' with respect to which the motions used in Newton's laws can be defined (or, at any rate, we need that rest frame itself, and it can't be replaced by any frame defined by some of the material bodies);
2. It is impossible to detect whether a body is at rest or in uniform motion.

These look contradictory. If we can't detect motion relative to absolute space, even indirectly, then how can that notion be required by physics? To see that there is no true contradiction, let's define an *inertial reference frame* (or just *inertial frame*) as any reference frame moving at constant velocity as measured by the rest frame—that is, for Newton, by absolute space. Bodies at rest with respect to some inertial frame are moving inertially—that is, in a straight line, at a constant speed—according to the rest frame, and indeed according to any other inertial frame.

The content of the relativity principle is now that physics can be done equally well using the standard of motion defined by any inertial frame—it doesn't have to be the rest frame. And, indeed, we can (mostly) see this just by looking at Newton's laws: the First Law says that a force-free body either remains at rest or moves at constant velocity, but if a body does so relative to one inertial frame, it does so relative to all inertial frames. And the Second Law relates acceleration to force—but because acceleration is *rate of change* of motion, the acceleration of a body is the same in any inertial frame.

What we require to do physics then, is *some inertial frame or other*. Given one such frame, we can construct indefinitely many other such frames—but we don't need to know which frame is the rest frame, and indeed, we can't know, because of the relativity principle.

Let's now return to the puzzle of how absolute motion can be detected (since we can't see the inertial frames, any more than we can see absolute space). Newton's example of the two globes demonstrates that we can determine whether a body is rotating by looking at the internal tensions within it. (Note that rotation isn't affected by changes of absolute velocity, so that if a body is rotating with respect to one inertial frame, it's rotating with respect to all of them.) His method works only because we presuppose Newtonian mechanics: we only know that there is tension in the string because the theory tells us that there should be. The generalization is: we find the absolute motions by looking at the relative motions and then asking ourselves: what would the motion of an inertial frame have to be, relative to all these bodies, in order that their motions relative to that frame satisfies Newton's laws? We know that process won't give us a unique choice of inertial frame, because of the relativity principle, but we can hope that for a sufficiently complex system, it will give us a unique choice *up to uniform velocity boosts*—or,

(a)

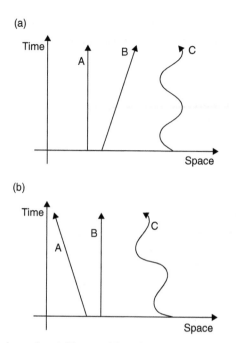

(b)

1. **Motion in two inertial frames: (a) motion in particle A's inertial frame; (b) motion in particle B's inertial frame.**

put another way, that it will give us the whole collection of inertial frames, even as it does not pick out one of them as preferred.

Figure 1 provides an illustration of this. In Figure 1(a), the motions of three particles A, B, C are shown with respect to an inertial frame in which A is at rest; B is moving, but in a straight line at constant speed; C is wobbling from side to side but not systematically going anywhere. Figure 1(b) redraws the same situation in an inertial frame where B is at rest; now it is A that is moving in a straight line at constant speed, while C continues to wobble, and now drifts to the left. The two inertial frames disagree as to which particles are at rest and which are at motion, but they agree that A and B are moving inertially and that C is not.

From this perspective, we can rewrite Newton's First Law:

> *Newton's First Law* (absolute-space version): An object on which no force acts will move in a straight line at constant speed relative to any inertial reference frame.

The concept of spacetime

Our new version of Newton's law makes no reference to 'absolute space' or to any 'rest frame', and so it is tempting to think that we can stop talking about absolute space altogether. But we need to be cautious. Recall the definition of an inertial frame: it's a frame moving at constant velocity *relative to absolute space*. If we excise absolute space from the theory, we seem to have lost the ability to say what an inertial frame actually *is*.

We could just accept that absolute space is required conceptually for our theory, even though it cannot be observed (even indirectly). In the language of Chapter 1, this would be a case of underdetermination (not of the correct theory, but of some fact about the world). Most philosophers regard this as a coherent possibility, but they tend to be uncomfortable about it—and physicists are usually even more uncomfortable. There is at least a strong pressure to find a way of thinking about the theory which removes the underdetermination.

To see how this might be done, recall the idea of absolute space as a background for physics: something which doesn't obey dynamical laws itself but defines the concepts that those dynamical laws require. Absolute space provides two separate notions of 'background structure' required for physics:

1. *Spatial geometry*: absolute space provides the standard that defines what the distance is between two bodies (which mathematicians call a 'metric'). This structure is *Euclidean geometry*, in the language of geometers, and it is just posited, with no suggestion (at least until Einstein) that it should have any dynamics of its own.
2. *Rest-frame structure*: absolute space provides a definition of rest and motion, enough to define the absolute velocity of any body.

A third piece of structure, also required, is given by absolute time, which (for Newton) is also a background notion:

3. *Temporal metric*: absolute time determines the elapsed time between any two events.

What the relativity principle tells us is that we don't require *quite* this much structure. We need spatial geometry; we need a temporal metric; we need *some* of the rest-frame structure, but only enough to tell us how much objects are accelerating (and which objects are not accelerating)—we don't need the full distinction between rest and motion.

Physicists have a very elegant way to think about these various structures, called *spacetime*. (The *word* 'spacetime' was not coined until the 20th century, as part of mathematician Herman Minkowski's exploration of Einstein's physics, but the *concept* is just as applicable to Newtonian mechanics.) Mathematically, spacetime is just a combination of space and time into one: if each point in space corresponds to *where* something might be, and each point in time corresponds to *when* something might occur, then a point in spacetime corresponds to both at once. That makes spacetime four dimensional: we need three numbers to say where something is in space (say, how far it is from a reference point in each of three directions) and one number to say when something is in time (say, the time at which it occurs, relative to some preferred moment like the year 0 CE), then we need four numbers to say both—and all mathematicians mean by a 'four-dimensional space' is a collection of things that are labelled (in an appropriately smooth way) by four numbers.

You don't need to be able to visualize things 'in four dimensions' to use the spacetime concept. In fact, here's a trade secret: hardly anyone, including professional physicists and mathematicians, really thinks in four dimensions—they just pretend the space they're thinking of is two or three dimensional and trust the maths to tell them when that pretence stops working. Let's try that trick here: pretend for now that space is only *two* dimensional, and think of it as a vast sheet. Given you can't see space, in fact, let's visualize it as a vast, thin sheet of Perspex. To get spacetime, stack many (strictly, infinitely many) copies of the sheet one on top of another: that whole infinite pile of spaces is spacetime. If you imagine taking a very thin needle and pushing it through the pile from bottom to top, so that it pierces each sheet once, then the path taken by the needle can be thought of as representing the trajectory of a particle: at each time, the hole in the sheet at that time represents where the particle is at that time.

We can now think of the various bits of background structure we just discussed as different bits of the structure of spacetime. The spatial geometry tells us the geometry of each of the copies of space: each of the sheets of Perspex, in our imagined model. Given two points on the same sheet, the spatial geometry tells us how far apart they are. The temporal metric tells us the separation in time of any two different copies of space.

What about the rest-frame structure? It can be thought of as telling us how the points of absolute space themselves evolve in time. Given two points in spacetime—two points on different copies of space—they either represent *the same* point of absolute space, or *different* points. This then allows us to define which particle trajectories—which paths through spacetime—represent inertial motion, and which represent accelerated motion. You can think of this as a preferred, special set of paths through spacetime: each one represents a particle at rest. The whole construction— spacetime, the spatial geometry, the temporal metric, and the rest-frame structure—is called *Newtonian spacetime*, in the normal terminology of philosophy of physics.

To do physics, though, we don't need the rest-frame structure, just the *inertial* structure, and we can represent this on spacetime by a whole family of preferred, special sets of paths in place of the single set of paths that defined the rest-frame structure. Each member of the family picks out one inertial frame—one collection of particles all moving at the same constant speed—but we don't regard one as preferred to the others. Doing so makes spacetime a little less structured: the resultant object is called *Galilean spacetime*, honouring Galileo's discovery of the relativity principle.

Does spacetime explain?

Newton's scientific case for absolute space was that it was required in order to define motion. From that point of view, spacetime might seem a friendly amendment to Newton's substantivalism: positing an enduring physical *space* was, in hindsight, overkill, because that gives us an unneeded notion of absolute rest, but positing (Galilean) *spacetime* is just what's required to equip physics with the background it needs. From this *spacetime substantivalist* perspective, spacetime is no mere mathematical construct or abstraction, but a *physical* background, replacing the separate notions of space and time. We can't detect it directly, but it's conceptually necessary to make sense of our physics.

There is a lot to be said for this approach—but there is also something a bit puzzling about it. In Newton's picture, it's at least reasonably clear what motion is: something moves if it occupies *different* points of space at different times; it is at rest if it stays at the *same* point of space as time passes. The spacetime replacement picture gets rid of the motion/rest distinction in favour of an inertial/non-inertial motion distinction, and then declares that something is moving inertially provided that the inertial structure of spacetime says it is.

But that starts to sound circular: 'motion is inertial if and only if it's one of the inertial-structure-preferred motions'. How do we understand what the inertial structure *is*, if it's not, by definition, the list of which motions are and aren't inertial? And if we can't understand it except by that definition, aren't we still left unsure what 'inertial motion' actually means?

There are basically two ways to respond to this concern. The first is to double down on the spacetime-substantivalist idea. We take the package of structures with which spacetime is equipped— spatial geometry, temporal metric, inertial structure—and just call it *spacetime geometry*. And we treat that spacetime geometry as a primitive feature of the world, irreducible to any facts about dynamics. The 'inertial structure' does indeed define a family of trajectories in spacetime, but that's just a basic, uninterpreted fact about the world. We then say that it's a *substantive law of physics* that, in fact, bodies on which forces don't act move along members of that family of trajectories. (A world in which force-free bodies moved some different way is conceptually possible, but the laws of physics tell us that our world isn't like that.) In this *geometry-first* approach to spacetime structure, it's a substantive fact about the world that it has the spacetime geometry it has, and a separate substantive fact that the geometry meshes with the dynamics in the way it does.

This first approach is probably the majority position among philosophers. But in making the logical separation it does between geometry and dynamics, it does leave us puzzled about what we actually *mean* when we say 'spacetime has such-and-such geometry'. Now, ultimately we have to take some concepts as primitive, unanalysed, and perhaps spacetime geometry is one such concept. But there is an alternative, which might be called *dynamics first.*

The dynamics-first approach (which I'll admit to being more sympathetic to myself) just takes as a definition that 'inertial frames' are frames in which force-free bodies move inertially. From that point of view, there isn't really any further analysis of the frames to be given: the laws of physics just make the claim that there are *some* frames with respect to which force-free bodies move in straight lines at constant speed, and then defines those as the inertial frames.

The difference between these approaches comes out most clearly when we think about the role of spacetime geometry. For the geometry-first approach, spacetime geometry *explains* various facts about the dynamics: because physical laws are always formulated with respect to some spacetime background, the geometry of that spacetime constrains what those laws might be. For the dynamics-first approach, spacetime geometry just *codifies* those facts about the dynamics: spacetime has the geometry it has because of the laws of physics, not vice versa. And as a consequence, it really matters for the geometry-first approach that spacetime is a physical thing; for the dynamics-first approach, it's as natural to think of it as a formal, mathematical tool (though it's contentious if that's really required by the approach, or even if it really makes sense).

Is anything at stake in this debate? I hope that the questions are fascinating in their own right—they concern deep matters of how the world is structured—but at a somewhat more practical level, they matter when we start to consider the evidence that the inertial structure of the world is not after all an unchanging background, but something that is affected, even determined, by matter and its dynamics. That evidence, and its implications, will be our focus for the last part of this chapter.

Inertia and gravity

Outer space is a 'zero-gravity' environment: astronauts in orbiting spacecraft are not pulled towards the floor, objects that they drop or throw just move in straight lines until they hit something, and everyday life in outer space is either strange and exhilarating (if you believe NASA publicity videos) or else really inconvenient (if you believe what astronauts actually say in hindsight about it).

Why is it a zero-gravity environment? Far too often one hears that it's because there is no gravity in space (presumably because it's far away from the Earth's gravitational pull). This is nonsense. The Earth has a radius of about 6,400 kilometres; the International Space Station orbits about 500 kilometres up; the gravitational pull of the Earth at 6,900 kilometres distance is scarcely less than at 6,400 kilometres. A better way to understand why astronauts don't experience gravity is that everything in an orbiting spacecraft is moving freely under gravity, at the same rate: the astronauts, their possessions, and the walls of the spacecraft itself. So gravity does indeed pull the astronaut towards the wall of the spacecraft, but it also pulls the wall of the spacecraft away from the astronaut, and to exactly the same degree. Hence the better name for 'zero gravity': *free fall*.

The reason this works is that gravity is what physicists call a 'universal' force: the acceleration it induces on a body is the same for small bodies and for large. (Contrast electrical forces, for instance: electrons are negatively charged, so they are attracted by the electric force of a big positively charged object; protons, being positively charged themselves, are repelled. But there is no analogue of this in gravity, no 'anti-gravitational' matter that is repelled by the gravitational field of the Earth, or even that feels it a bit less.)

But universal forces pose a puzzle for our understanding of inertia. Suppose you are an astronaut on the first interstellar space

Philosophy of Physics

mission, and in all the excitement you fall asleep. On waking, you find yourself weightless, so your spaceship has at least left the Earth's surface, and the engines aren't currently turned on—but you don't know if the ship is (a) still in Earth's orbit, ready for a last conversation with Mission Control; (b) partway to Jupiter Base, where you're going to refuel; or (c) in the space between the stars. Without looking out of the window or asking someone, you have no possible way of knowing, because in each case you are in free fall, and so you can't do any experiment to detect the gravitational field.

This should sound familiar. It has very much the same structure as Galileo's thought experiment of the ship: just as uniform *velocity* is undetectable, so are uniform *external gravitational forces*. But recall the lesson of that thought experiment: because absolute velocity boosts can't be detected, we can't need a notion of absolute rest to formulate physics. The analogous argument is: if we can't detect whether a system is uniformly accelerating under some external gravitational field, it can't be necessary to formulate the physics of that system to distinguish between the absence or presence of an external gravitational field.

What is the alternative? This: just as Newton's laws hold not just with respect to the *rest frame* but to any *inertial frame*, so they hold with respect to any frame that is *falling freely under gravity*.

This is a radical rethinking of how inertial motion works in physics. First, it means a force—like gravity—that is universal is not a force at all: forces, according to Newton's laws, cause deviations from inertial motion, and free motion under gravity just is inertial motion. Instead, gravity defines the inertial frames, and then the other forces in the world—the non-gravitational interactions—tell us how objects accelerate relative to those inertial frames.

Second, it breaks the idea that inertial structure is *background* structure. If the inertial frames are determined by the gravitational interactions, and thus by the distribution of masses, then inertial structure is no longer a fixed background against which physics plays out.

Finally, it tells us that inertial structure is *local*. Thus far, an inertial reference frame has been something that can be defined for the whole Universe at once—but if inertial structure is determined by gravitational effects of matter, and if those effects vary from place to place, then inertial structure likewise must vary from place to place. The notion of a single collection of inertial frames is replaced by a patchwork, one collection for each little region of spacetime. This in turn requires us to set rules for how adjacent collections of frames are related to one another; these rules are what physicists mean by *spacetime curvature*.

(Einstein is generally credited with this remarkable insight about gravity: he considered what it would look like to be in an elevator that was falling freely through its shaft, and concluded that it would be as if there was no gravity at all. Famously, he called this realization 'the happiest moment of my life'. But Newton must have understood it to some degree: the Earth–Moon system, for instance, is constantly accelerating as it orbits the Sun, yet Newton knew that he could apply his physics to the Earth and the Moon as if the Sun were absent.)

Is this change in our conception of inertia *required*, though? Arguably not. We could carry on claiming that inertial structure is absolute and unaffected by gravity—but the cost is that the 'inertial' structure becomes quite undetectable, rather as the absolute rest frame was undetectable given the relativity principle. This is a live possibility on the geometry-first way of thinking about spacetime, where there is no conceptual relation between spacetime structure and the motions of bodies. From that perspective, these insights about free fall at most suggest that we should look for a different spacetime structure (just as the unobservability of absolute rest was a good reason to get rid of absolute space, but didn't actually force us to do so). By contrast, on the dynamics-first approach, the realization that inertial motions are determined by gravity is directly a discovery about spacetime structure. The distinction is subtle, to be sure, and there is plenty of room for disagreement, but the case of gravity rather forcefully makes the case for understanding the geometry of spacetime as codifying, not as logically independent from, the notions of inertia used in the laws of physics.

Most of the philosophy in this chapter was developed in the past few decades—some of it is cutting-edge work—but the physics is centuries old. In Chapter 3 we will catch up with modern physics and consider the conceptual questions raised by the theory of relativity—but we will see that most of the insights we have acquired here transfer over to that newer, and stranger, theory.

Chapter 3
Relativity and its philosophy

In popular culture, the central idea of the theory of relativity—the source of its name, indeed—is Einstein's deep insight that *motion is relative*. Armed with that insight (it is said) Einstein went on to revolutionize our understanding of space and time, overthrowing the ideas that had reigned since Newton.

We have already seen that this cannot be the whole truth—the principle of relativity (though not the name) dates back at least to the 17th century. The original of relativity theory, as we will see, actually comes from the apparent incompatibility of the relativity principle with other, apparently secure, discoveries of late 19th-century physics, and from Einstein's insight that compatibility could be restored at the cost of changing how we think of space and time. In this chapter, we will see how this trick is done, explore some of its paradoxical consequences, and reconsider the puzzles and controversies about motion and space that we explored in Chapter 2 from this new and unintuitive perspective.

Problems for the relativity principle: light as a wave

Try to solve this elementary maths problem:

A fighter aircraft, on the ground, can fire bullets at a speed of 340 metres per second. The aircraft has a top speed of

260 metres per second. If it fires its bullets forward while flying at that top speed, how fast are they going?

The obvious answer is of course 340 + 260 = 600 metres per second, and (if we neglect air resistance) that is the physically correct answer too.

But here's a structurally similar problem with a different answer:

> The sound waves emitted by a fighter aircraft's engines travel at 340 metres per second. The aircraft has a top speed of 260 metres per second. You are in front of the aircraft and hear its engines; how fast are the sound waves going when you hear them?

The right answer *here* isn't 340 + 260 = 600 metres per second: it's just 340 metres per second. Sound waves travel at the speed of sound, no matter how fast the source of the sound might be going.

Why is there a difference? Bullets are flung out of a gun: they pick up the speed of the gun. But sound waves propagate in the air: the speed they travel at is fixed by the physics of the air and doesn't depend on how fast the source of the sound is going. (The *frequency* of the sound waves depends on the speed of the source—this is the famous Doppler effect, familiar to anyone who has heard the wail of a police siren change pitch as it passes them—but the *speed* does not.)

A corollary is that how fast the sound is going relative to *you* depends on how fast you are going relative to the *air*, whereas for the bullet stream, that hardly matters at all: only your speed relative to the *source* of the bullets matters. The lesson generalizes to any wave-like phenomena (sound waves in solid objects, water waves on the sea, etc.): the speed of a wave is fixed relative to the medium that the wave travels in, and does not depend on the speed of the source of the wave. But *you* will only observe the wave speed to be constant if you are stationary with respect to the medium. If you are moving in the medium, you will observe that the waves have different speeds depending on which way they are going.

Here's a different way to put it, drawing on the ideas of Chapter 2. The speed of a wave is defined with respect to a special choice of inertial frame: the frame in which the medium that is waving is stationary. If you measure the speed of the wave in any other inertial frame, you'll get a different answer, and an answer that depends on the direction that the wave is going. On the other hand, the speed of a bullet is defined with respect to the inertial frame in which the source of the bullet is at rest.

Does any of this violate the principle of relativity? Not really, any more than the fact that objects fall down violates the idea that there is no preferred direction in space. In each case, some material thing is breaking the underlying symmetry of the laws. But the physics of that thing itself still obeys the principle of relativity: if the air itself is in motion, for instance, then the speed of sound is measured relative to that moving frame.

But this argument does rather rely on the fact that sound waves in the Earth's atmosphere are a local, even parochial phenomenon. If, impossibly, the whole Universe was filled with air, and if that air never displayed wind or other local movement, then it would be impossible to get outside the local, symmetry-broken environment. In that situation, it would become less clear whether the relativity principle really holds.

The Universe is not filled with air, but it is filled with light. And while it isn't obvious or intuitive that light is a wave, by the beginning of the 20th century there was a great deal of evidence that it was. The great physicist James Clerk Maxwell, building on a half-century of important work, had established that a changing magnetic field could create an electric field, that a changing electric field could create a magnetic field, and that the whole process—electric to magnetic to electric to magnetic to...—would be self-sustaining, and would travel through space at light speed. The idea that light just *was* that wave received spectacular experimental confirmation, most notably in the creation, transmission, and reception of radio waves in the late 19th century.

If light is a wave, that seems to imply a medium in which it travels—a medium which the physicists of the day called *aether*, undetectable to the senses but essential for the understanding of light. The speed of light would be independent of the speed of a source, but would be defined relative to the aether rest frame.

That raises two puzzles: one conceptual, one practical. The conceptual puzzle is: doesn't this conflict with the principle of relativity? After all, the aether isn't local or parochial like the Earth's atmosphere: since light can exist anywhere, the aether must fill space. And an undetectable, space-filling medium relative to which the motion of light is defined starts to sound a lot like an absolute rest frame, a lot like Newton's absolute space—concepts which we got rid of (didn't we?) back in Chapter 2.

The practical puzzle is: if there is, after all, an absolute rest frame that defines the speed of light, how fast are we moving relative to it? The Earth is rotating around the Sun, and the Sun is in turn rotating around the centre of the Galaxy; there is no obvious reason to expect the aether frame to coincide with the motion of the Earth. And—unlike in the Newtonian case—it looks as if it ought to be possible to *measure* the Earth's speed relative to the rest frame, just by measuring the speed of light in different directions. (Recall: you get the same speed in all directions for a wave only if you are stationary with respect to the medium that is waving—and if you get a different speed in different directions, that lets you measure how fast you're moving relative to that medium.)

Now, these are not easy experiments to do. Light moves at a staggering 300 million metres per second (nearly 700 million miles per hour); meanwhile, the Earth's velocity relative to the Sun is around 30,000 metres per second, and the Sun's velocity relative to the centre of the Galaxy is around 150 metres per second. So we are talking about very subtle predictions, very tiny changes in the measured speed of light. But it can be done, and the result is quite clear (and, indeed, it was pretty clear even in the early 20th century): if there is an aether frame, Earth is not moving relative to it. Put another way: in the Earth's reference frame, the speed of light does not depend on its source.

It was already a bit disturbing to have to accept the aether at all, and effectively to give up on the principle of relativity. It is even more disturbing to find that something as parochial as *the Earth* fixes the aether frame. One can come up with reasons, of course—physicists in the early 20th century talked of *aether drag*, whereby the aether is pulled along by massive bodies—and those reasons might have led to fruitful research programmes. But there is still a feeling that we may be missing something.

The rise of relativity

Let's review the dilemma. If light is a wave, then there must be a medium in which it travels. That medium effectively defines a rest frame, in conflict with the principle of relativity; and we can detect that rest frame by looking for the frame in which the speed of light is independent of the source. So we either have to give up on the principle of relativity (and thus abandon our discoveries about space and inertia) or give up on the wave theory of light (and thus abandon all those remarkable experimental predictions).

In ordinary human life, clashes between irreconcilable principles are sadly common, and compromise or prioritization are the only ways forward. But the history of physics tells us that Nature really doesn't like to compromise. When two deep principles of physics appear to be in conflict, it very often turns out that what must be abandoned is not either principle, but some hitherto-unquestioned background assumption.

So, following Albert Einstein in 1905, let's ask: what if the relativity principle still holds, *and* the speed of light is independent of the speed of the source? Putting those together tells us that the speed of light is independent of the speed of the source in *all* reference frames. That looks impossible—but let's look a bit more closely. Specifically, suppose I fire a pulse of light at 300 million metres per second, and you chase it at, say, 200 million metres per second. Then the light is getting ahead of you by only 100 million metres per second, so (surely?) that's how fast you'll measure it as moving. (And if you had run in the opposite direction, you'd measure it as moving at 500 million metres per second).

But to measure the speed of light in your reference frame, you'll need to bring along some measuring tape—or, better, a nice rigid measuring rod—and a good watch. *You* will measure the light you're chasing as moving at 100 million metres per second only if your measuring rod, and your stopwatch, agree with mine. And what Einstein realized was that this is a substantive physical assumption, not just a truism. He showed that we can hold on to the relativity principle *and* hold on to the idea of the speed of light being independent of the source—provided we are willing to let not just our standard of *motion* but our standards of spatial and temporal geometry vary from one inertial frame to another. That is: in Einstein's theory of relativity, notions like spatial distance and temporal duration, cease to have an absolute, frame-independent meaning. What is 'relative' in relativity is not just motion, but time and space.

What all that means *mathematically* is, a century later, beyond all dispute. And equally beyond dispute is that the mathematics works: Einstein's modified laws on how reference frames are related are at the core of modern astrophysics and particle physics. But what it means *conceptually*—how to understand it—is another matter, and we will spend most of the rest of the chapter exploring it.

Time dilation and the twin paradox

I just described the content of relativity in terms of the relation between different observers in different reference frames. But its central physical implications can be described for a single reference frame. The two main ones are:

- *Time dilation*—In a moving system, all the physical processes slow down relative to those same processes in a stationary system. In particular, a clock that kept good time when stationary will run slow if it is in motion.
- *Length contraction*—A moving object shrinks (in the direction of motion) compared to the same object when stationary. In particular, a stationary measuring rod will be shorter in motion than it was at rest.

(There is a third, more subtle, effect, which we will discuss shortly.)

It is because of length contraction and time dilation that I can accept that a moving observer speaks truly when they say 'I measured the speed of light and got 300 million metres per second', even while I myself got the same result for my own measurement. The two are not contradictory because—as *I* describe it—your measurements were actually made with slowed clocks and shrunken measuring rods, and so by my lights they need to be rescaled in order to be accurate. The two raise closely analogous issues; for reasons of simplicity and space, I will focus on time dilation.

Time dilation is a *directly observable* prediction of relativity. Because the speed of light is so fast, it is most pronounced in the behaviour of sub-atomic particles (the only things we can realistically accelerate to speeds close to light speed). Many of these particles are unstable—they decay into other particles—and they have characteristic decay times, which makes them clocks of a sort. Relativity predicts that these decay times slow down, the faster the particles are moving, and exactly this is observed, both in human-built particle accelerators and in the natural experiments that occur when fast-moving cosmic rays hit the Earth's atmosphere. The predicted time dilation can be significant—a factor of ten in cosmic-ray experiments, for example—and experiments exactly reproduce those predictions. Large time dilations for massive objects are harder to produce, but modern atomic clocks are so accurate that they can measure even the tiny time dilations caused by putting the clock in an airliner. Again, the experiments reproduce the predictions.

Yet there is an apparent contradiction in the very idea of time dilation. Moving clocks run slowly, I said—but motion is relative. If you are moving rapidly relative to me, I predict that your clock runs slowly. But *I* am moving rapidly relative to you, and—according to the relativity principle—you predict that it is my clock that slows down. That starts to sound close to a contradiction: how can two clocks each run more slowly than the other? If A is twice as slow as B, and B in turn is twice as slow as A, doesn't that make A four times as slow as itself?

This *clock paradox* can be sharpened into one of the most famous thought experiments in physics, the *twin paradox*. Two twins fall out badly, and one of them heads off, in a rocket and in a huff, at (say) 80 per cent of light speed. Five years later (as measured on the Earth), he regrets his anger and heads back at the same speed (Figure 2).

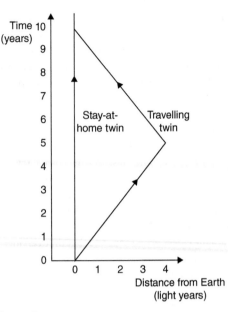

2. The twin paradox.

The stay-at-home twin reasons thus: my twin has been travelling for ten years, at 80 per cent of light speed: relativity predicts that time will have slowed down for him. His clock, his smartphone, and his ageing process will all have advanced only (as it happens) six years. So when he returns, he will be younger. (For vastly lower speeds, and vastly smaller effects, those experiments where clocks are carried around the world by airliners effectively have this form—so the observed slowing down of the moving clock on the airliner can be thought of as an experimental confirmation of this predicted slowdown.)

But wait (the stay-at-home twin goes on). This whole time, *I* have been travelling at 80 per cent of light speed relative to *him*. So *he* is just as entitled to conclude that I will be younger than him. But we can't both be younger than each other! Just as with the clock paradox, the twins thought experiments appears to make time dilation not just problematic or puzzling, but actually inconsistent.

The inconsistency in each case turns on the apparent symmetry between two accounts of what is going on. In the clock paradox, the moving clock runs slowly with respect to the stationary one, and then the principle of relativity seems to say that by the same token the stationary clock runs slowly with respect to the moving one. In the twin paradox, the moving twin ages less than the stationary one, and then the principle of relativity seems to say that by the same token the stationary twin ages less than the moving one. The beginning—though not the end—of understanding what is going on is seeing how this apparent symmetry can be broken.

Making time dilation consistent

Breaking the symmetry in the twin paradox is fairly simple. There is a significant difference between the two twins: the stay-at-home twin spends all her time moving *at the same velocity*, whereas the moving twin turns around half way through his voyage. So the second half of his voyage takes place at very high velocity relative to the first half.

That suffices to remove the *contradiction*—there is not after all a perfect symmetry between the twins' situations, so it isn't logically impossible for one twin to end up younger than the other. But the moving twin can still reason: my stay-at-home sister is moving rapidly compared to me; so her clocks are running slow; so she will be younger when I see her. We still lack an *explanation* of what is wrong with this argument, or why the moving twin's large change of velocity leads to overall time dilation. But we have at least established that there is nothing inconsistent in the idea that it might do so.

51

(Sometimes—even among physicists who ought to know better—one hears the idea that it is the *acceleration* that causes the time dilation. That's somewhere between very misleading and flatly wrong: the acceleration is required only in the sense that the moving twin has to accelerate in order to do one part of his journey at a very different velocity to the second part. And if the stay-at-home twin, feeling bored, decides to accelerate up to 80 per cent of light speed for a few seconds, then to turn round and fly at 80 per cent of light speed in the other direction for a few seconds, and then finally stop again, she will have done as much acceleration in total as her twin, yet it makes essentially no difference to the age gap between the twins.)

The clock paradox is subtler—but also takes us closer to an actual explanation of what is going on. Recall the structure: the moving clock runs slow compared to the stationary clock. But how is that actually to be measured? Here's a protocol:

1. Get *two* good watches and give one to a friend.

2. Stand in the path of the moving clock (or the spaceship it's on). Get your friend to stand further down the path.

3. At the instant the moving clock passes you, note down (i) the time on your watch, (ii) the time on the moving clock.

4. Your friend does the same when the clock passes them.

5. Get together with your friend and compare notes. You can now work out both how much time elapsed according to your watches, and how much time elapsed for the moving clock. The ratio of the two is the time dilation.

The most important thing to note here is that we are really comparing not two, but three clocks: the moving clock is running slow compared to a *pair* of stationary clocks. The relativity principle then tells us that, conversely, a stationary clock will run slow as measured by a *pair* of moving clocks. But we can't combine these two and conclude, as the clock paradox threatened, that a single stationary clock runs slow compared to itself.

But inconsistency still seems to loom in the background. Imagine *two* pairs of clocks: one pair stationary, one pair moving. It looks as if we can conclude that each pair runs slowly compared to the other—and that, too, looks contradictory. To see why in fact there is no contradiction even here, we need to look a little more carefully at the measurement protocol. And doing so reveals that time dilation and length contraction are not the only novel features of special relativity. There is a third: *relativity of simultaneity*.

Spreading time through space

One of the key ideas in our protocol to measure time dilation is that we can find how much time passed in our reference frame during the flight of the moving clock by comparing my watch's reading when it starts its flight, to my friend's watch reading when it finishes that flight. And the same would be true if we were interested in measuring the speed of the clock (or any other moving body): the difference between my watch reading and my friend's is the travel time, and the speed is then the distance between the two of us divided by that travel time.

This strategy requires our two watches to be synchronized—to be reading the same time. If my watch is running five minutes slow, the measurement protocol will be unreliable. So: how can we do this? There are two natural options:

(i) My friend and I make sure our watches are synchronized when we meet up in person and agree the protocol, and before we walk to our respective spots.
(ii) After my friend is in place, I phone them up and we synchronize the watches over the phone.

In ordinary circumstances, either protocol works fine. (The first is what criminals and special forces soldiers do—at least in movies—before the heist or the hostage rescue; the second is how your phone or computer updates its time settings over the internet.) But both are pretty problematic in relativity. For the first: time dilation means that moving clocks slow down, so even if our clocks are synchronized when we meet up, they might not stay synchronized. For the second: telephone signals (and any other signal we might try) move at a finite speed, so we need to allow for the finite travel time of our synchronization signals. But we can't do that, because it requires us to know how fast the signal is going—and we need to synchronize watches before we can measure signal speed!

The first time one encounters this problem, it can seem technical or even pedantic. Isn't there a simple solution, an uncontentiously straightforward way to synchronize watches? But there isn't. (Try it.) The upshot is that measuring the speed of moving bodies (or measuring time dilation) actually requires three components: reliable measuring rods, reliable clocks in multiple locations, and a *synchrony rule* to decide how to coordinate the clocks—what the philosopher Harvey Brown calls a rule for 'spreading time through space'. More than one rule is possible, and different rules will give different speeds for a moving body.

You should now be worried about the foundation of relativity. Didn't I say, only a few pages ago, that the theory is founded on (1) the relativity principle, and (2) the axiom that the speed of light doesn't depend on the speed of its source? If the speed of light depends on a choice of synchrony rule, that axiom looks ill-defined.

There is an ingenious way around that problem. Measuring the time something takes to travel from A to B requires two clocks (one at A, one at B) and so needs a synchrony rule—but measuring the time it takes to go from A to B *and back again* requires only one, at A. So you can measure the 'two-way speed of light' all by yourself, with a metre rod, a stopwatch, a flashlight, and a mirror, as follows:

1. Stand at one end of the metre rod, and put the mirror at the other.
2. Point the flashlight at the mirror. Turn it on, and at the same instant, start your watch.
3. As soon as you see the flashlight in the mirror (which means that the light has travelled to the mirror and back again), stop the watch.
4. You now know how long it took the light to travel 2 metres, to the mirror and back. The two-way speed is then [2 metres] divided by [elapsed time on stopwatch].

(This requires good reflexes: you will need to stop the watch about six nanoseconds after starting it!)

We can now state the axiom thus: 'the *two-way* speed of light is independent of the speed of the source and of the direction in which the light is emitted'. And this fact not only serves as a basis for relativity, but let us define a very natural synchrony rule, the *Einstein synchrony rule*: we should synchronize our clocks so that the *one-way* speed of light is also independent of the speed of the source and the direction of emission. That choice is guaranteed to be well-defined by the assumed constancy of the two-way speed.

Here's how the rule works in practice. The two-way speed of light is 300,000 kilometres per second so let's define a 'light second' as 300,000 kilometres. You are standing, let's say, 3 million kilometres from me—ten light seconds away. I send you a signal, saying 'my watch is reading 12:00:00'. You now set your watch to 12:00:10, so that according to our two watches, light took ten seconds to cross those ten light seconds, and the one-way speed of light is also 300,000 kilometres per second.

Here's a different way to describe that same method. Suppose that I send a signal to you and you immediately bounce it back to me. When the signal gets back to me, my watch reads 12:00:20 (independent of any synchrony rule). You need to set your watch so that the time it reads at the moment of bouncing is exactly *half way* between the time on my watch when the signal left (12:00:00) and the time when it got back (12:00:20)—that is, you should set it to 12:00:10. That half-way choice guarantees that we record the light as moving at the same speed on the way out as on the way back.

Physics uses the Einstein synchrony rule almost exclusively, and the formulae for time dilation (and actually length contraction too) assume that rule. So most of the measurements of time dilation I discussed previously—cosmic rays, particles in accelerators, etc.—should really be thought of as picking up time dilation *according to Einstein's rule*. But not all measurements should be thought of that way. In the twin-paradox thought experiment—and in the clock-on-an-airliner experiment—one clock goes out and back again, and then is found to have run slow compared to the other clock. We don't need any synchrony rule to state or test that claim—it might be thought of as a measurement of 'two-way time dilation'.

The Einstein rule sounds natural, even obvious, but it has one very non-obvious conclusion: judgements of when two events happen at the same time depend on which inertial frame you are using. To see this, let's suppose that you are ten light seconds due east of me, and that somewhere west of us there is a bright flash of light. I see the flash at 12:00:00, and at 12:00:10 you reflect it back to me with a mirror. So if I glance at my watch at 12:00:10 then your reflection and my glancing are simultaneous.

Simultaneous for *me*, that is. Let's suppose that Alice and Bob are travelling due East at half the speed of light, directly away from the source of the flash. As I glance at my watch, Alice shoots past

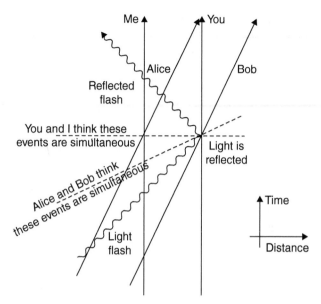

3. Relativity of simultaneity.

me, and as you receive my signal, Bob shoots past you at the same
speed. Bob and Alice can use that same flash of light to
synchronize *their* clocks (see Figure 3).

However: I saw the flash at 12:00:00, but at that time, Alice
hadn't reached me. She must have seen the flash earlier, as it
passed her on the way to me. And when you bounce the flash of
light back towards us, Alice will receive the rebounding signal
before me, because in the intervening time she's crossed some of
the distance to Bob. So, the half-way point between Alice sending
a signal and it bouncing back to her occurs earlier than the
half-way point between *me* sending a signal and getting it
bounced back. Which means that Alice and I disagree about
which times are simultaneous with the signals reaching you
and Bob.

So—at least according to the Einstein rule—scientists in two frames that are moving relative to each other don't just disagree on the lengths of bodies or the rate at which time flows—they disagree about which events happen at the same time. This *relativity of simultaneity* is quite sweeping in its implications—given any two events such that light cannot pass between them, there will be some choice of frame where those events happen at the same time, some choice of frame where the first happens before the second, and some choice of frame where the second happens before the first.

(This, incidentally, is one way of seeing why relativity strongly suggests that it is impossible to travel faster than light: any faster-than-light signal is a backwards-through-time signal with respect to some reference frame.)

And this (finally) resolves the contradiction in the clock paradox. As judged in any one frame, a moving pair of clocks not only are running slow, but are wrongly synchronized. To go from the stationary clocks to the moving clocks and back again, we have to correct (twice) for change in synchronization as well as for time dilation—and (it turns out) when this is done there is after all no contradiction in the claim that each pair of clocks is running slow as measured by the other pair.

Conventionality of simultaneity

Einstein synchrony makes the clock paradox consistent—but it has very strange consequences. We seem to be saying that what is happening *right now* depends on my velocity—if I change velocity, distant happenings move from the past into future or vice versa. To drive home the strangeness, imagine that right now your beloved sibling is getting married in the Andromeda Galaxy, two million light years away. You pace back and forward, worrying whether everything will go well—but during your pacing *forward* the Einstein synchrony rule means that the wedding won't happen for hours, whereas as you pace *backwards* the wedding finished yesterday.

But is Einstein's rule *right?* It is certainly possible to define different conventions for synchronizing clocks: we could decide to make the one-way speed of light different in different directions, say, or we could even just pick the frame we like most and say that clocks in *all* frames must agree with *that* frame as to which events are simultaneous. But Einstein's rule has major advantages over all of these: some of them require us to pick out a preferred reference frame; others pick out a preferred direction in space; more pragmatically, the equations of physics are far simpler if we adopt Einstein synchrony than any of these rivals. So we can certainly say that Einstein synchrony is the right *practical* choice.

Saying that it is the true choice, and that the other choices are *wrong* (and not just ill-advised) is another matter. Some analogies may help. In timekeeping, the normal rule on Earth is to split the world into twenty-four zones with twenty-four different choices for when noon is, chosen in each case so that 12:00:00 occurs roughly when the Sun is directly above. An alternative rule would be to require that all clocks, everywhere on Earth, keep time with Greenwich Mean Time. We don't usually use that alternative rule, because it's inconvenient for noon to occur in the middle of the night in some parts of the world, but that doesn't make it wrong, just less useful. (Indeed, in some circumstances—military operations, for instance—it turns out to be more useful, with the benefits of a shared time outweighing the costs of that time coming apart from Sun and Moon.) And the reason that neither rule is right is that there is nothing objective to be right *about*— independent of our conventions, there just is no fact of that matter about what *the* time is.

A second example: graph paper is nearly always produced with the lines at right angles, and coordinate systems for maps likewise nearly always have the axes at right angles. This choice has many advantages, some obvious and some more subtle. Pythagoras's theorem, for instance—the result that the square of the distance between two points on a plane equals the square of their separation on the x-axis plus the square of their separation on the y-axis—holds only if the axes are at right angles. For almost all purposes, it will be *less useful* for you to have graph paper where the lines are at some other angle. But the graph paper is not *wrong*, because again, there just is no fact of the matter as to what the angle *really* is between the axes—those axes are just our own choice as to how to coordinatize space.

What is at stake in this case is whether the question 'what is happening *right now* at some faraway place?' has an objectively correct answer (in which case there would be an objectively correct synchrony rule, and one could then make a good case that it was the Einstein rule), or whether it is simply a matter of convention what the answer is (in which case we can understand the Einstein synchrony rule—or the Einstein synchrony *convention*, as it is more commonly called—as a very sensible choice for that convention, but not the objectively true choice).

(There is a slight subtlety about the word 'conventional' here. Some conventions are *pure* conventions: there really is no reason at all why we should use 'dog' and not 'chien', 'canus', or 'Hund' to refer to dogs. By contrast, right-angle axes and location-relative standards for when noon is are *good* conventions: it is objectively sensible to use them in most situations. But the right standard to assess them is sensible-vs-daft, not true-vs-false.)

One philosophical argument for a unique answer to the question comes from the philosophy of time and from the widespread idea that there is something fundamentally different between past and future. The past (on this way of thinking) is fixed; the future is yet to come and so is open; the present is on the edge of transition between future and past; as time flows, the past constantly grows and the future recedes. If that really is the deep nature of time, it seems to point to a sharp and non-conventional notion of simultaneity.

That theory of time has long been controversial in philosophy. It has been argued that the flow of time is just an incoherent notion: flow is a thing that happens *in* time, and so not something that makes sense for time itself. (A popular slogan among critics: 'how fast does time flow—one second per second?') But more importantly for our purposes, it conflicts sharply with the idea that simultaneity is relative (which is forced on us by the Einstein synchrony convention, and indeed by any remotely plausible 'objectively right' convention). If I can move an event from past to future and back again just by pacing up and down, and if the events that are past for me are future for you just because we are moving at different speeds, it is hard to see how there can be a fundamental distinction between the nature of past and future. An objective notion of simultaneity seems to require a frame-independent, absolute notion of simultaneity: once we accept that simultaneity is relative, it is only a small step further to the view that simultaneity is conventional. (But this is one place where philosophers continue to defend many different positions; defenders of objective simultaneity—or of the objective flow of time—have by no means given up.)

Minkowski spacetime

To see more clearly what conventionality of simultaneity really means, let's return to the spacetime concept and update it for relativity. Recall that we visualized spacetime (in Chapter 2) by suppressing one dimension of space (so that space could be thought of as a thin Perspex sheet) and then built spacetime from a pile of those sheets, one for each instant of time. That spacetime—which we called 'Galilean spacetime'—comes equipped with three bits of structure: spatial geometry (defining distances and angles on each sheet), temporal geometry (defining the time separation between sheets), and inertial structure (picking out the inertial frames with respect to which force-free matter moves in straight lines). In the light of relativity we can now see that is a further, tacitly given *simultaneity structure*: two spacetime points are simultaneous if and only if they are on the same sheet.

The easiest way to visualize the spacetime of special relativity—normally called *Minkowski spacetime*, after the mathematician who proposed it soon after Einstein's original work—is to imagine heating the stack of sheets up until they melt together, leaving us with a block of Perspex but erasing the individual sheets. In special relativity's spacetime, events are not organized into families all of which happen at the same time, there is simply a four-dimensional collection of those events.

Of course, that collection is far from structureless. In Newtonian spacetime (i.e., spacetime before the lesson of the relativity principle), for any two events we could say separately how far apart they are in space and in time. In Galilean spacetime, we can still say how far apart two events are in time, but their space separation is undefined (except relative to some arbitrary inertial frame) unless they are simultaneous. In Minkowski spacetime, neither spatial *nor* temporal distance between events really makes sense unless we fix an inertial frame. They are replaced by a unified concept: *spacetime distance* or, as it is usually called, the *interval*. As Hermann Minkowski puts it, in one of the more famous passages in physics,

> Henceforth space by itself, and time by itself, are doomed to fade away into mere shadows, and only a kind of union of the two will preserve an independent reality.

What does this 'interval' actually represent, physically? When two events represent stages of the life of the same persisting object—or, more generally, when a fast-enough-moving particle could travel

Philosophy of Physics

62

between them—the interval between them is the flight time that would be measured on a clock that travelled between them in a straight line. (Events like this are called 'timelike separated'.) When instead the events cannot be connected by a moving body or even a light ray ('spacelike separated'), there will be some inertial frame in which they are simultaneous—and then the interval is the ordinary spatial distance between them, as measured in that frame.

There is a strikingly simple formula for spacetime distance. Pick any inertial frame, and use Einstein synchrony to synchronize the clocks for that frame. Then given events A and B, we have

$$\left(\text{Interval A} - \text{B}\right)^2 = \left(\text{Time distance A} - \text{B}\right)^2 - \left(\text{Space distance A} - \text{B}\right)^2$$

where the time distance is measured in seconds and the space distance in light seconds, or the time distance in years and the space distance in light years, etc. You might notice a striking similarity with the Pythagorean formula we discussed earlier—but there is a crucial minus sign. The interval becomes 0 when the time distance equals the space distance—that is, it is zero when A and B are connected by a light ray, travelling at one light second per second. (And if the space distance is greater than the time, we have to reverse the minus signs in order to get out a meaningful formula.)

Although this formula was defined with respect to one inertial frame, it gives the same result in any inertial frame: indeed, one way to construct the equations of relativity is to deduce the frame-independence of the spacetime distance from the relativity principle, the postulate about the two-way speed of light, and Einstein's synchrony rule.

It is vital to notice that while the background structures of Minkowski spacetime are importantly different from those we saw in Newtonian physics, nonetheless they *are* background structures in the same sense as before, unchanging and essential for the physics of material bodies. In particular, a notion of inertial structure is just as essential in relativistic as in non-relativistic physics. That said, the inertial structure is not independent of the spacetime distance; indeed (though the details will not matter), mathematically we can recover inertial facts entirely from spacetime-distance facts. Whether that means *physically* that inertial structure is secondary to spacetime-distance structure is another matter, as we will see as we return to our central question: how to understand time dilation.

Geometrical and dynamical explanations of time dilation

Let's consider the twin paradox again, from the spacetime viewpoint. The travelling twin goes out at 80 per cent of light speed, travelling four light years in five years; then he returns. We can identify three key events in the story (see Figure 4):

A: The travelling twin leaves Earth
B: The travelling twin turns around
C: The travelling twin returns to Earth.

In spacetime terms, the stay-at-home twin goes from A to C in a straight line. The travelling twin goes from A to B in a straight line and from B to C in a straight line. The time recorded by the stay-at-home twin's clock is then equal to the interval from A to C; for the travelling twin, it is the interval from A to B plus the spacetime distance from B to C.

In ordinary space, the shortest path between two points is a straight line. But the minus sign inverts this: in Minkowski spacetime, straight lines are the longest routes between two points of spacetime (at least for timelike separated points). And so the

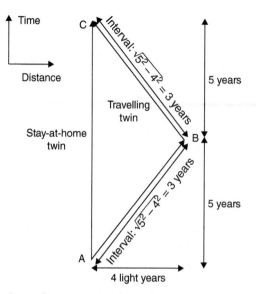

4. The twin paradox: geometrical description.

geometry of Minkowski spacetime tells us that the time elapsed for the stay-at-home twin will be longer.

Indeed, we can go further, and use the definition of the interval to calculate the actual difference. The interval from A to C, in years, is the square root of $(10^2 - 0^2)$, or just 10—as we'd expect, the stay-at-home twin has a ten-year wait. The interval from A to B is the square root of $(5^2 - 4^2)$, or 3—for the travelling twin, only three years pass before he turns round. And by symmetry, the interval from B to C is again 3, so that the total time measured by the moving twin is six years—time passed only 60 per cent as quickly for the moving as for the stationary twin.

So: the geometry of Minkowski spacetime predicts that the travelling twin will record less elapsed time (because the longest spacetime path between two points is a straight line). It makes clear the asymmetry between the twins that resolves the twin paradox (because the moving twin turned round, his spacetime path is bent, whereas the stay-at-home twin's path is not, and that affects the spacetime distance). And it even gives us a ready way to calculate the difference in times for the two twins.

But does it *explain* why the moving twin records less total time? The question is really a repeat of the geometry-first vs dynamics-first question we encountered in Chapter 2: is the geometry of spacetime an explanation of the physical phenomena (in particular of the difference between the twins' ages) or simply a codification of those phenomena? If the former, we can reasonably take the description I gave above as a true explanation. But if the latter, then the spacetime geometry just encodes the facts about the twins; it does not tell us why those facts obtain.

To see what a dynamics-first answer to that 'why' question looks like, let's return to the general notion of time dilation. We've seen already that time dilation is only well-defined relative to a choice of inertial frame: in each frame, moving clocks run slow. From the geometry-first perspective, this frame-dependence makes time dilation suspect: notions like 'moving' and 'slow' only make sense relative to a frame, can't be understood in terms of the invariant structure of spacetime, and so shouldn't really be playing a role in explanation. From this perspective, it is really twin-paradox effects that are the true content of time dilation: clock slowdown, outside the context of a clock that goes away and then returns, is mathematically and calculationally useful as a concept but not really fundamental.

From the dynamics-first perspective, the fact that clock slowdown requires an inertial frame to define it doesn't make it unreal: inertial frames are the basis of how we do physics, and it is only natural for dynamical explanations to be carried out in one frame or another. And in any such frame, when we say 'a moving clock runs slow'—or, for that matter, 'a moving rod shrinks'—we mean that the physical processes inside the rod—the interatomic bonds that hold the rod together and define its length, the periodic processes that count time inside the clock—are different for matter in motion than for the same matter when stationary. The electric field of a moving charge, for instance, shrinks in the direction of motion—according to the laws of electromagnetism—compared to the field of a stationary charge. Ordinary matter is held together by electric fields, so if those fields are altered by motion, then it is only to be expected that the shape of the matter will be altered.

Despite this concrete electromagnetic example, we don't actually have to study the detailed microphysics of our clocks and rods in order to predict time dilation and length contraction. All we need to know is that the forces that hold them together, collectively, define a physics which satisfies the relativity principle and delivers a velocity-independent speed of light. That defines a constraint on the form of the laws tight enough to guarantee that if they can describe rigid bodies and good clocks at all, those bodies and clocks will, when in motion, conform to the relativity principle.

On this explanation—in contrast with the geometry-first case—the moving twin has recorded a slower time because he was moving throughout the period, and in doing so his clocks slowed down due to their internal physics being affected by motion. How does this explanation avoid the paradoxical aspect of the twin paradox?—after all, from the point of view of the moving twin, it's the stay-at-home twin who is moving!

The answer is that, while dynamical explanations can be given for any inertial frame, the frame that moves at the same speed as the moving twin is not an inertial frame, because the twin turns round. We can identify *two* relevant inertial frames: the 'outbound frame', co-moving with the twin until turnaround, and the 'inbound frame', co-moving with the twin after turnaround. (See Figure 5.)

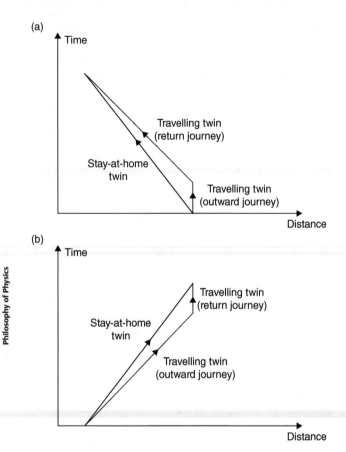

(a)

Time

Travelling twin
(return journey)

Stay-at-home
twin

Travelling twin
(outward journey)

Distance

(b)

Time

Travelling twin
(return journey)

Stay-at-home
twin

Travelling twin
(outward journey)

Distance

5. **The twin paradox: inertial-frame description.**

In the outbound frame (Figure 5(a)), the stay-at-home twin is
moving at 80 per cent of light speed throughout. Until
turnaround, the 'moving' twin is stationary—but after he turns
around, he is moving very rapidly (40/41 of light speed, in fact),
even more rapidly than the stay-at-home twin.

So, described in the outbound frame: the stay-at-home twin's clock runs slow compared to a stationary clock. The moving twin's clock initially doesn't run slow, but after turnaround, it runs even more slowly than the stay-at-home twin's, so much so as to record overall less time than the stay-at-home twin. Conversely, in the inbound frame (Figure 5(b)) the moving twin's clock initially runs very slowly compared to a stationary clock, then runs at its normal speed, while the stay-at-home twin's clock runs somewhat slowly all the while. In each case, we have a physical, dynamical explanation for the twin's discrepant time measurements, in terms of the effect of motion on their physical processes—but the explanation is different in each inertial frame.

To sum up: On the geometry-first approach to time dilation:

- The twin-paradox time gap is a real physical effect, explained by the fact that the moving twin takes a longer path through spacetime.
- Talk of clocks 'slowing down', by contrast, is a reference-frame dependent notion that doesn't have an invariant spacetime description and should be avoided.
- The paradox of time dilation is dissolved once it is noticed that the moving twin's path is bent whereas the stay-at-home twin's path is straight.
- Genuine explanation of spacetime phenomena should ideally be done with reference to spacetime and should avoid reference to inertial frames.

On the dynamics-first approach:

- Moving clocks really do run slow, as a consequence of the physical processes that determine their timekeeping and the effect of motion on those physical processes.

- The twin paradox can be explained, in any inertial frame, by this slowdown of the physical processes of bodies moving with respect to that frame.

- The paradox of time dilation is dissolved once it is noticed that the stay-at-home twin is stationary with respect to one unchanging inertial frame, whereas there is no inertial frame with respect to which the moving twin is stationary throughout his voyage.

- It is fine for an explanation of spacetime phenomena to be given with respect to a given inertial frame, because dynamics is defined with respect to those frames, though the relativity principle means that an explanation ought equally well to be available for any other inertial frame.

In this book, I can only present this dichotomy, not resolve it—and of course it is more complicated and nuanced than I can fully convey here, with intermediate and synthesized versions available. For all that the mathematics of special relativity is old and well-understood, these interpretative questions about the theory remain open, with philosophers and physicists alike writing and thinking about them in very different ways.

Epilogue: general relativity

Einstein's *general* theory of relativity, developed a decade or so after the special theory, is a central topic in modern physics and in modern philosophy of spacetime, but its technical complexity puts the deep conceptual puzzles it poses beyond the scope of this book. Still, we can build on what we have learned in these last two chapters to understand at least the basic idea of the theory.

Specifically: general relativity is a merger of the discoveries about inertia and gravitation that we discussed at the end of Chapter 2, with the discoveries about inertial structure and the workings of the relativity principle that comprise special relativity. Recall: the universal nature of gravitation tells us that inertial structure is not given as an unchanging, Universe-wide background, but rather is determined locally, by the distribution of matter. The effects of gravity are then understood in terms of the curvature of spacetime—that is, the ways in which local inertial frames at one place relate to those at nearby places.

In Newtonian gravity, the physics for those local inertial frames is Newton's physics. General relativity simply arises—in a mathematically very natural, almost unique, way—when we keep the insight about gravity and just replace the Newtonian concept of an inertial frame with the one we get from *special* relativity.

(There is, however, one important change that comes about when special relativity and gravity are merged. Just as conventionality of simultaneity meant that space and time structure could not be fully separated, it means also that we cannot have a theory of *spacetime* curvature without it also being a theory of curved *space*. What that means, and what its ramifications are for the questions of this chapter and the last, is beyond the scope of this book.)

In understanding general relativity in this way, we also get some insight into just what Einstein's genius was. It was not simply that he came up with new ideas—ideas are cheap, frankly. It was that he understood features of our existing physics—whether the relativity principle or the relation between gravity and inertia— more deeply than his predecessors. Einstein did not *replace* Newton's theory of gravity with a theory in which gravity was about spacetime curvature—he realized that Newton's theory of gravity was *already* a theory of spacetime curvature, and then worked out how such a theory would look if its local notion of inertia was appropriately changed.

Or at least, that is how we might think about general relativity if we focus on the idea of inertial structure and inertial frames. There is a different way to understand the theory: start with Minkowski spacetime, and then ask how that theory could be changed to make the spacetime interval a dynamical entity which depends, and reacts back, on the matter distribution of spacetime. That way of thinking about the theory leads to the same mathematics, but a quite different understanding of the physics—the deep interpretative questions of this chapter and the last extend into general relativity, even as they shift and grow more complex in the beautiful and subtle context of our best contemporary theory of gravity.

Though there is far more to say about the philosophy of space and time, we now change focus. Chapter 4 considers not any one specific theory of physics, but the relation between multiple theories at multiple levels of description: the domain of statistical mechanics.

Chapter 4
Reduction and irreversibility

Modern chemistry might be said to start with the 'periodic table'—the discovery, in the 19th century, that the chemical elements can be organized into families, with their position in the family allowing some of their core properties to be predicted. Modern *quantum* chemistry began with the momentous discoveries, in the early 20th century, that the structure of those families, and indeed the chemical properties of the elements, could be predicted from the physics of the electrons, protons, and neutrons which, we learned, comprised those elements.

These discoveries make up one of the clearest examples of *reduction* in science—where a larger system is explained in terms of its smaller constituents, and where ideas in the field of science that deals with that larger system turn out to be analysable via the ideas of the field that deals with the smaller one—in this case, atoms are explained in terms of subatomic particles, and (a part of) atomic chemistry is explained in terms of—as philosophers say, is *reduced to*—physics. Reduction*ism*—at least at first pass—is the idea that reductions like this are the template for all relations between scientific theories, and that ultimately the concepts of higher-level, larger-scale scientific theories will reduce to lower-level, smaller-scale theories, and ultimately everything will be grounded in physics.

Enthusiasm for reductionism ran high in the first half of the 20th century, and sometimes went hand-in-hand with a rather dismissive attitude by physicists towards the higher-level 'special sciences'—exemplified by the (possibly apocryphal) comment attributed to the physicist Ernest Rutherford that 'all science is either physics or stamp collecting'. More recent times have seen a backlash against reductionism, driven by observations about the complexity and autonomy of the special sciences and the apparent irrelevance of the details of physics to their character, as well as lingering concerns about (supposedly) intrinsically irreducible features of human experience—pain, consciousness, and the like. Nowadays it is common to say that higher-level sciences are 'emergent from', not 'reducible to' physics, and the exact relation between these notions of emergence and reduction remains contested.

All of which might seem to have little to do with physics. If physics is the science that other things reduce *to*—if physics, whatever its relation to the other sciences, is the field of science that studies matter on the smallest and most fundamental scales—then considerations of emergence and reduction at first glance look irrelevant to physics itself. But the reality is that only a small part of physics is really concerned with studying 'the most fundamental scales'. Most systems physicists study—atomic nuclei, metals, plasmas, climates, galaxies—are themselves complicated, larger-scale systems whose relations to lower-level physics are complex and indirect. And so understanding reduction—and, more generally, the relation between theories at different levels—is actually a key conceptual issue for physics, and for philosophers of physics. We will see that even within physics, reduction is a subtle matter, requiring entirely new concepts beyond the basic dynamical laws of physics—concepts of *irreversibility* and *probability*. Both entered physics in the 19th century, but both remain contested to this day.

On the plurality of physics

It's common to talk about 'the' laws of physics, as if at any time there is some single set of principles, some single group of equations, that encompass the content of physics. Indeed, I indulged in this temptation in Chapters 2–3, referring to Newton's laws as if they fully described pre-relativistic physics, only to be replaced by the laws of relativity a century or so ago. But the reality is that physics has dozens—hundreds—of different laws, hundreds of different systems of equations, describing different systems at different scales of description.

For example, along with the equations of Newtonian mechanics (useful for describing, e.g., how bodies move under gravity):

- The *Navier–Stokes equation* describes liquids and gases, at the level where they can be considered continuous and infinitely divisible;

- The *Boltzmann equation* describes dilute gases, at a somewhat finer grain;

- *Euler's equations* describe the tumbling of solid bodies;

- *Maxwell's equations* describe how electromagnetic fields evolve.

What is more, in most cases these equations can describe many different systems. To apply the law requires certain real numbers—like the viscosity (stickiness) of a fluid, or the mass of a planet—and those numbers vary from one application to another. (In a sense 'the same' equations describe the Earth–Moon system and the Sun–Jupiter system, just with different assignments of masses to the two planets.) So really, we should think about physics not as giving a single description of the world, but as giving a plethora of descriptions, of different systems, at different scales.

Yet these descriptions are not independent of one another. Often, we can describe one and the same system in more or less detail, and then try to understand how the two descriptions are related. It's helpful to have a concrete example in mind here, so let's talk about gases: specifically, the air in the room in which I'm typing this. There are, very roughly, 1,000 trillion trillion air molecules in the room—10^{27} in scientific notation. In principle, knowing the positions and the velocities of all 10^{27} molecules right now (that's now 6×10^{27} numbers—the x, y, and z components of position and velocity for each molecule) would suffice to predict what they will do in the future.

In practice, that is basically never how we study gases. Here's an alternative description: instead of giving all 6×10^{27} numbers, we give a coarse-grained description of the gas—say by breaking the room into cells, 1 cubic millimetre on a side, and giving the pressure, density, temperature, and average velocity of the gas in each cube. That's still a lot of information—very roughly, I need a hundred million, or 10^8, numbers to tell you that much about the gas—but 10^8 is a lot smaller than 10^{27}. At this level of description, I'm no longer discussing the gas as a collection of particles, but as a fairly continuous fluid. (And, with luck, the pattern of variation of those numbers across the room is fairly smooth, so I can summarize them in a much more succinct way.)

The name of the game, now, is to look for an *autonomous* description of the gas at this level of description. What I mean by 'autonomous' is that if I want to know the fluid-level description of the gas in (say) five minutes' time, I can deduce it from the fluid-level description now. That is: to know how those 10^8 numbers change in time, I don't need the initial values of the 6×10^{27} numbers that specify the position and velocity of each particle, but only the 10^8 numbers that give the current fluid-level description.

And we do seem to have that higher-level description, at least in this case—it's the Navier–Stokes equation, briefly mentioned above. (Different coarse-grained descriptions give different higher-level equations—the Boltzmann equation can be obtained this way, for instance.) These equations are themselves not simple to work with—and, after all, 10^8 is still a large number—but in the right circumstances we can either solve them explicitly or at any rate learn general things about them. As a particularly important example, these equations predict that in due course the gas will evolve into a uniform state, where pressure, density, and temperature are constant across the room—and that when it reaches that state, the average temperature, pressure, and density will be related by a simple equation called the *ideal gas law*. At this point we are concerned not with 6×10^{27} numbers, not with 10^8, but only with three.

Looking for autonomous higher-level descriptions of this kind is the task of *statistical mechanics*—so-called because giving higher-level descriptions of systems tends to involve averaging over the statistical properties of their smaller constituents. It has been enormously successful since its inception in the late 1800s, yet it raises deep puzzles which persist even now. The first and most straightforward is simply: why do we want these 'autonomous higher-level descriptions' anyway? If we can describe the gas in all its microscopic detail, why settle for a partial account?

One popular line—especially in physics textbooks—is that statistical mechanics is necessary because of our cognitive and experimental limitations. That is: we don't actually have the ability to measure exactly where every single particle is, and even if we did have that ability, it's too difficult to solve the equations to calculate how they change over time. If we were smarter and had better equipment, we could dispense with statistical mechanics entirely.

A related argument is that we are only *interested* in certain features of a system. We don't care about the exact positions and velocities of all the particles, only about coarse-grained summaries of those positions and velocities—and so statistical mechanics lets us extract the information we need about the features of the system that we actually want to study, without getting distracted by trivia.

But there are reasons to be sceptical that this is the whole story. After all, it just seems to be a *fact* about the world, a fact that we would like to explain and understand, that fluids obey the Navier–Stokes equation, or that the ideal gas law holds for gases that have settled down into uniform states—and, furthermore, these facts were discovered long before they were analysed through statistical mechanics. Even if we did have the calculational and experimental ability to predict, for any given initial state, exactly how a system like the air in my room would evolve, that alone does not seem to tell us why the autonomous high-level description exists, or what it is. (An exact micro-level description of the air in the room could at most tell us that for *this* configuration of molecules, the ideal gas law holds; it could not tell us whether to expect it to hold for configurations in general.)

Furthermore, the role of human interest seems overstated here. To be honest, I'm *not* all that interested in the air in this room, beyond the basic requirements for breathability—yet despite my lamentable lack of curiosity, it's still objectively true that there are autonomous higher-level descriptions of that air. Conversely, there are individual features of the world that I'm extremely interested in—the sales figures for this book, for instance. But for most of them, I can't understand how they change over time without knowing a lot about other, much less interesting features—there is no autonomous dynamics for my sales figures! Whatever it is about the world that makes some coarse-grained features describable autonomously and susceptible to the methods of statistical mechanics, and other features less so, it isn't anything as simple as 'what humans care about'.

What we see here is a tension between two different conceptions of statistical mechanics. On the *inferential conception*, statistical mechanics is a tool of inference, used to study complex systems in the face of our partial knowledge and variable interest. On the *dynamical conception*, statistical mechanics is about understanding and discovering the various objectively correct higher-level descriptions of complex phenomena, and learning how they connect together. In this conception of statistical mechanics, it is simply the tool that physics uses to study emergence. For the reasons just given, I'm more sympathetic to the dynamical conception, and this chapter is mostly written from that perspective, but both provide distinctive insights, and the reality may be more complex than a simple binary choice between them.

But whatever perspective we choose, there are deep puzzles with how the higher-level descriptions we actually use could *ever* be extracted from lower-level physics. Indeed, there are apparently plausible arguments that it is impossible to do so, that the higher-level descriptions have features—two features in particular, *irreversibility* and *probability*—which in principle could not be extracted from the microscopic level. In the rest of the chapter, we'll see what these features are, and how the apparent contradictions they lead to might after all be avoided.

Reversibility and irreversibility

Imagine watching a speeded-up video of the Earth rotating, or the moons and planets of the solar system orbiting one another; could you tell if the video was played backwards? Well, yes, perhaps, if you remembered a bit about the phases of the moon or recalled that the Sun rises in the east and sets in the west—but it requires thought. There's nothing *immediately wrong* with the video if it's played backwards, nothing that makes it *obvious* that something is wrong. The movements of the planets look basically the same forwards and backwards.

Now do the same trick with a video of a heap of sand collapsing, or of milk mixing into coffee, or (speeded up) of ice melting, or living things growing and decaying. Now it's completely obvious whether the video is being run backwards: there are immediate features of all these processes which tell us which way they are being played.

The technical term for all this is *reversibility*. The motions of the planets are reversible; the collapsing of heaps and the melting of ice and so forth are irreversible. And the fact that some dynamical processes are reversible and some are irreversible creates profound difficulties for statistical mechanics.

To see how, let's get a bit more precise as to what reversibility means. As a starting point, let's think about dynamical laws in physics as devices for predicting the future: if you input the present state of the system to which the laws apply, the device spits out what its state will be in one second's time, or one hour's time. For the air in the room, for instance, if you input the positions and velocities of all of the particles now, the laws determine what the positions and velocities will be at whatever future time you're interested in—always supposing that no outside influence intervenes. For the solar system, once you've specified the present position and velocity of the planets, that suffices to determine what they will be in the future: for instance, we can and do use those laws to work out the dates of future solar eclipses. The equations of motion for milk mixing into coffee are intractably difficult to solve, but, in principle, if you input the initial data about which bit of the cup has coffee in and which bit milk, they likewise let you predict those facts for all future times. All the laws we have considered so far are *deterministic*, which means that for any given input they provide a *unique* output at each future time.

The essence of reversibility is that reversible laws can equally well be used to predict the past—to *retro*dict, as philosophers sometimes say. The motion of the planets is reversible: the present data about the planets can be used just as easily to determine past and future eclipses. (This is important in ancient history, for instance: knowing exactly when the eclipses happened can quite precisely date a source, if that source mentions an eclipse.) The mixing of milk or the melting of ice is irreversible: information is lost in the process, so that many early states of the coffee cup or the cold drink are compatible with the same later state. And so insofar as we can use the dynamics of irreversible systems to learn things about their past state, it has to be done indirectly and via additional background assumptions.

Two related notions help us see the significance of reversibility. A system is *recurrent* if it endlessly repeats itself, so that if we input some initial data, the system's data at some later time will match those initial data. A system has an *attractor region* if there is some state (or some collection of states that doesn't contain all the states) such that whatever the system's initial state, eventually it ends up at the attractor state(s) and then stays there indefinitely. (In statistical mechanics, attractors are very often called 'equilibrium states', and the process of evolving towards one is called 'equilibration'.)

Roughly speaking: systems are recurrent if and only if they are reversible; they have attractors if and only if they are irreversible. So we can take 'recurrent' as an alternative way of saying 'reversible', and 'has attractors' as an alternative way of saying 'irreversible'. The argument, in essence, is that if a system is not recurrent then there are some states that the system will leave and never return to; the set of all states except those is then an attractor region, and the system irreversibly enters that attractor. (There are some caveats here: notably, the system we're studying has to be confined to some finite region, say a box or a room.)

We can now see why irreversibility poses a problem for statistical mechanics. The microscale physics of systems—like the physics of the individual particles in the dilute gas—appears reversible, while the physics at larger scales—like the physics of the gas at the fluid scale—is normally irreversible. And (on the face of it) that makes it impossible for the latter to be a coarse-graining of the former.

Let's use the notions of recurrence and attractors to spell this out. If the microscopic system is recurrent, then any state of the system eventually comes back to where it started. But that means that any *coarse-grained* description of the system eventually comes back to where it started, too. In other words, if a system is recurrent, any coarse-graining of that system must also be recurrent.

We can do it the other way around, too. Suppose that the coarse-grained, higher-level description has an attractor region, and suppose that the system starts outside the attractor region. Then it will enter the attractor region *and never come back out.* But in that case, the microscopic system cannot be recurrent. However we look at it: irreversible large-scale physics could not *possibly* be emergent from reversible microscopic physics.

Or rather: it could not be emergent unless some extra ingredient is added. What these arguments tell us is that coarse-graining can't be the whole story: there must be some additional assumptions or conditions added to the microscopic, reversible physics in order for irreversibility to turn up at a coarse-grained level. But to understand what it might be, we need to consider the second feature of high-level physics that is missing from the low-level description: probability.

Probability in statistical mechanics

I noted in the last section that all the laws we were considering were *deterministic*: they gave unique predictions for the future, given the present. But not all laws in physics have that form. The physics of sub-atomic particles seems to display genuine randomness, as we will discuss in more detail in Chapters 5–6. But even outside that strange realm, there are phenomena in nature that defy deterministic description—at least at the level that we study them.

A classic example is Brownian motion. If a pollen grain is suspended in a fluid, and studied through a microscope, it will be seen to jitter around, apparently at random, actually because of its constant collisions with water molecules. (The jitters don't reflect individual collisions, but the average of many collisions.) It's not possible to write down deterministic equations for those jitters—even perfect information about the grain's initial position and velocity is insufficient to determine what it will do next. But it *is* possible to write down a so-called 'stochastic equation' for the grain—an equation which does not say what the grain will definitely do next, but does say how probable each possible jitter is. (It may say, for instance, that the grain is equally likely to jump in each direction, that the average length of a jump is 10 microns, and that longer or shorter jumps have such-and-such probability of happening.) Stochastic equations are perfectly testable (by running many repeats of the observation and collecting the statistics) and offer useful, informative descriptions of many systems.

Now, it is tempting to say that the apparent randomness in the pollen grain's behaviour is not *real* randomness. If we knew the exact, microscopically precise, information about all of the water molecules, couldn't we predict *deterministically* what the pollen grain will do? Well, perhaps. But in doing so, we have given up on the idea of giving an *autonomous* description for the pollen grain's dynamics—which, recall, is our goal in statistical mechanics. And we have given up not because no such description is to be found, but because of our refusal to allow that autonomous description to include probabilities. Turning it around: Brownian motion illustrates that the story we tell about how large-scale physics emerges from small-scale microphysics must have room in it not just for irreversibility, but also for probability.

As a mathematical matter, it's easy enough to see how this can be done. Since the microscopic dynamics knows nothing of probability, the only place to add it is in the description of the initial state of the system. There will normally be a great many microscopic states compatible with any given coarse-grained description of the system (remember the gas, where the microscopic description needed 10^{27} numbers while the coarse-grained description needed 10^8—in that case, the coarse-grained description leaves those numbers almost completely unspecified, corresponding to a plethora of microscopic states giving rise to the same macroscopic physics—each macroscopic state corresponds to $10^{27}/10^8 = 10^{19}$ microscopic states). If we say of the microscopic system not merely that it is in *some* such compatible state, but that it has a certain *probability* of being in each compatible state, then we have a route by which probabilities can enter our macroscopic physics.

This is not just a speculation. Physicists have a special choice of probabilities—called the *uniform probability measure*—which roughly speaking says that each microscopic state compatible with the macroscopic description of a system has the same probability. (It actually says something a little more subtle, since strictly speaking there are infinitely many such states.) If one starts with the uniform probability measure, then it's actually not too difficult to calculate the equations for Brownian motion, probabilities and all. We could summarize this, schematically, by the equation

Deterministic microscopic physics + uniform probability measure
$\rightarrow$ Stochastic macroscopic physics

This introduction of probability also offers an apparent route around the problem of irreversibility that we encountered in the previous section. We have seen that it is impossible for *every* microscopic state compatible with a given coarse-grained description to evolve according to some irreversible large-scale physics for that system. But this doesn't rule out the possibility that *the vast majority* of such microscopic states (as measured by the uniform probability measure) might evolve according to that large-scale physics—or, at least, that they might do so *for a very long time*. (Recurrence means that they cannot do so forever.) And, indeed, this is the orthodox physicists' answer to the irreversibility problem:

> yes, it's not *certain* that the system will obey the irreversible laws—and it can't do it *forever*—but it is *almost certain* that it will obey those laws, and for a very long time.

And this is not just an in-principle possibility. The actual method used by physicists to construct higher-level equations from lower-level physics is exactly to impose the uniform probability measure and then deduce the coarse-grained dynamics—and this method *works*, in the very pragmatic sense that the equations that are derived match experiment. As another schematic equation, we might write

Reversible microscopic physics + uniform probability measure
→ Irreversible macroscopic physics, almost certainly

These insights—that statistical mechanics requires probability to be added to microscopic physics, and that doing so allows us to reproduce irreversibility in practice—are the conceptual core of modern statistical mechanics. They have been profoundly successful, underpinning a huge body of empirically fruitful science. But they do not suffice to make the conceptual status of statistical mechanics unproblematic, for two reasons: the concept of probability is mysterious, and in any case—as a matter of logic—it cannot, alone, explain the emergence of irreversibility.

What are statistical probabilities?

Let's understand the uniform probability measure, loosely, as the statement that each microscopic state compatible with a given macroscopic description is equally probable. What does that actually mean?

Here's one thing it could mean:

> I don't know what the actual state is, so I think each one is equally likely. The uniform probability measure expresses my ignorance of the true state.

This approach is a natural fit to what I called the 'inferential conception' of statistical mechanics—the idea that statistical mechanics is a set of tools to let us make inferences about complicated systems given the limitations imposed by our ignorance. And it has the virtue of clarity: it's fairly widely accepted, in physics and philosophy alike, that probabilities can be used to quantify how sure or unsure of something we are.

However, it shares—and makes more explicit—the main disadvantage of the inferential conception, which is that it seems ill-fitted to explain the objective, high-level regularities that we observe in nature. In the case of Brownian motion, for instance, it seems to be just an objective fact that the particle is as likely to jump one way as another—and we can test that fact by collecting the statistics for many pollen grains. We could have known this—indeed, we did know this—before we were confident that individual water molecules exist, let alone what their detailed physics is, so it is hard to see how those observed statistics can have much to do with our ignorance of the precise positions and velocities of the water molecules.

(There is also a more technical problem for this approach. As I briefly mentioned above, the uniform probability measure is only metaphorically described as 'each state is equally likely'. In reality there are infinitely many states compatible with any macroscopic description, and it is a much more subtle and contested matter to say what the 'right' way is to express our ignorance of the true state.)

But if this 'ignorance interpretation' of statistical-mechanical probabilities fails to explain the actually observed behaviour of the systems we study, it must be admitted that the alternatives are not obvious. One natural possibility is:

> By probability, here, I really just mean frequency. There are lots of systems like this one; the uniform probability measure is a measure of the relative frequency of microstates, across all those systems.

This *frequency interpretation* of the probabilities is often what one finds in the textbooks; again, though, it is not fully satisfactory. For one thing, frequencies like this don't seem suited to explain why *this system right here* displays the behaviour it does. For another thing, there are lots of things 'system like this' might mean, and it is uncomfortable at best if our explanation of central statistical-mechanical phenomena like irreversibility ends up depending on questions of classification like this.

There is (much) more to say in defence of both the ignorance and the frequency interpretations—yet there is no generally accepted version of either, and they do not exhaust the possibilities. (My own, fairly minority, view is that we need to find some way of explicitly adding probabilities into the physics of individual systems, and that the ultimate origin of those probabilities is quantum-mechanical.) The interpretation of the probabilities in statistical mechanics is one of its central philosophical puzzles— second only to the problem of irreversibility, to which I now return.

Reversibility in, reversibility out

Computer programmers have a saying: 'Garbage in, garbage out'. It means that however clever a program might be, ultimately it has to work with the input it gets—if that input is faulty, that fault will be transferred to the output. Philosophers of statistical mechanics live by a similar saying: reversibility in, reversibility out. Which is to say: if your higher-level emergent physics is irreversible, and you claim to have derived that irreversible physics from some reversible lower-level physics by means of some assumptions, then either you have cheated or else one or more of those assumptions built in the presumption of irreversibility.

The point is important enough to be worth spelling out. An irreversible process makes a fundamental distinction between the past and the future: just looking at the dynamical equations suffices to tell you which is which. A reversible process does not: we can regard either direction in time as 'past' or 'future', with equal validity—at least as far as the mathematics is concerned. So if we derived the former from the latter, something must have been added during the derivation to break the past/future symmetry.

At this point, there is a temptation to ask: what could *possibly* break the symmetry in this way? (At which point speculation can run wild.) But there is a better way to approach the question: given that physicists actually have a method for deriving irreversible higher-level equations based on adding the uniform probability measure to the microscopic physics, where does *that method* break the symmetry?

At least in principle, that question has a simple answer. Recall: the uniform probability measure is the assumption that each microscopic state compatible with a system's coarse-grained description is equally likely. A small fraction of those microscopic states will not display the expected coarse-grained dynamics, but the vast majority will—so that we can be almost certain that those dynamics will in fact be displayed. But we can then ask: if the uniform probability measure is imposed on the initial state of the system, will it automatically continue to hold for later states? And the answer is that it will not: if, for instance, we are studying the system for some fixed time, so that it makes sense to speak of its final state as well as its initial state, then the probability distribution for the final state will be nothing like the uniform probability measure.

We can also see this another way. Suppose we ignore the claim that the system's 'initial' state really is its first state, and evolve it backwards (we can do this, remember, because the microscopic dynamics are reversible). By symmetry, we should expect that coarse-graining that backwards evolution will give a time-reversed version of the irreversible macrodynamics—a backwards-in-time dynamics that allows us to predict the *past* coarse-grained description from its present value. Put vividly: if we apply the uniform probability measure at some state of the coffee where the milk is partially mixed in, and then evolve backwards, we will 'predict' (retrodict, really) that in the past, the coffee and milk were even more mixed. On this approach, the history of the coffee cup is that it begins with coffee and milk fully mixed, goes through a brief period in which they unmix, and then starts mixing again. The instant of least mixing is the instant at which we imposed the uniform probability measure.

What this tells us is that, if the vast majority of states are ones that will evolve into the future according to the irreversible dynamics, then, equally, the vast majority of states are ones that will evolve into the past according to the time-reversed irreversible dynamics. As philosopher David Albert puts it, the vast majority of states are 'in the process of turning around'. And that means in turn that if a system's initial state is selected according to the uniform probability measure, the probabilities of final states are concentrated on a tiny minority of states that obeyed the ordinary irreversible dynamics (not their time-reverse) in the past.

The upshot of all this is that in applying the uniform probability measure, we are selecting a preferred instant of time. Statistical mechanics only predicts irreversibility if we insist that this preferred instant is the very first instant at which we consider the system. It works perfectly to predict the system's later behaviour, but gives wildly wrong predictions for how the system evolved before that instant.

So, what justifies this? There are two very different sorts of answer available, corresponding to the two very different conceptions of statistical mechanics which we have discussed.

The roots of irreversibility

Recall: on the inferential conception of statistical mechanics, the idea of the project is to give us tools to study systems when we have only partial information about them—only coarse-grained information, say. From that point of view, insofar as we want to make predictions about a system's *future*, the uniform probability measure might well be the best we can do. But to make predictions about a system's *past*, we can do much better, for the vital reason that we already have records, information, knowledge about the past. To try to retrodict the past using only the uniform probability measure is to use only a tiny fraction of the information we have—so no wonder it gives poor results.

Similarly, suppose we actually try to set up a system and then watch how it evolves—we prepare it so it has a certain coarse-grained description, but we lack the experimental precision to fix its exact microstate. So we're ignorant about that microstate, and the uniform probability measure is a natural way to represent that ignorance. And if we want to predict the system's state at later times, the methods of statistical mechanics are the best we can do.

But it would obviously be folly to try to predict the system's state at earlier times using those methods. They assume that the system is evolving under its own dynamics, without outside influence, and we know that that's false in its past, because our own preparation process was that 'outside influence'.

The central idea here is that the distinction between past and future in statistical mechanics is just a consequence of the distinction imposed by our own nature as experimenters and agents. Our memories and our abilities to intervene in the world determine a direction in time, and the irreversibility of statistical mechanics just follows from that direction.

This is an elegant approach, and is quite popular among some physicists, especially those with an interest in information theory. But it pays a high price for that elegance: in taking our human perception of the directedness of time as an *input*, it rules out any attempt to explain where that direction itself comes from. In particular, on pain of circularity it cannot (it would seem) appeal to the asymmetries and irreversibilities of macroscopic physics to explain *why* we, as physical systems, in fact have the capacity to remember the past and influence the future, but not vice versa. Furthermore, it seems to lack the resources to explain why, as a matter of plain fact, various physical processes throughout the Universe just display irreversibility, even when we have nothing to do with them. The melting of snow, the eruption of volcanoes, the birth and death of stars—all occur outside our control, and yet all manifestly obey irreversible laws.

What is the alternative? Recall: the uniform probability measure is an initial condition on a system: it correctly reproduces irreversible dynamics into the future, but not into the past. The earlier in time we impose it, for any given system, the longer the period in which the methods of statistical mechanics work. The limiting case is as simple as it is dramatic: impose the condition on the Universe as a whole, at the beginning of time.

The general name for a condition of this sort is a 'Past Hypothesis': a specification of the microscopic details of the Universe just after the Big Bang. The details of what that hypothesis should be are somewhat contested (when one looks at the details, there are more possibilities than simply imposing the uniform probability measure) but the basic idea is common to all: irreversibility requires an initial condition as well as the reversible microdynamics, and holds only after that condition is imposed; if irreversibility is an objective agent-independent feature of the world, then we need to treat the imposition of that condition as a fact about the world and not just about our way of interacting with the world. And to impose the condition consistently, we are led back and back until we end up imposing it at the creation of the Universe.

There is, to be sure, something startling, even bizarre, about the idea that the observed irreversibilities we see here and now have their origin in cosmology. Yet if we have a dynamical conception of statistical mechanics—if we indeed want to see the theory as an account of objectively correct emergent higher-level physics—its logic is difficult to avoid.

There is one, apparently innocent, assumption throughout this chapter: that it makes sense to suppose that a microscopic system actually has one state or another—that its constituent parts really do have some velocities and some positions, even if we do not know them. That assumption is called into question by *quantum mechanics*, without doubt the strangest theory in modern physics. In the last two chapters of the book, I turn to the philosophical puzzles of this remarkable subject.

Chapter 5
Mysteries of the quantum

A 'theory', in physics, can be many things, from the highly specific to the utterly general. At one extreme are the theories (sometimes called 'models') which describe particular, concrete systems, fully specifying—at least, on some level of description—what properties the system has and how it behaves. The mechanics of the planets of the Solar System is like this: it says that there are eight planets and the Sun, that their masses are such-and-such, and that they move in such-and-such a way. At a slightly more general level is the abstract *framework* of Newtonian celestial mechanics, which encompasses not just our solar system, but any system of bodies moving under gravity in the regime where relativity does not matter.

We can zoom out further. Special relativity does not describe any specific system or set of interactions: rather, it is a framework to encompass any theory whose inertial structure is that which we derive from the relativity principle and the light postulate (or, if you prefer, whose natural spacetime setting is Minkowski spacetime), whether that theory concerns particles in an accelerator or matter on a far larger scale. It is contrasted not with specific theories like Newtonian celestial mechanics but with alternative frameworks for writing such theories, like the framework described by Galilean spacetime. Similarly, the insight of the equivalence principle—that gravitation is the localization of inertial structure—is a framework for theories of gravity (or, rather, multiple frameworks, depending on whether the inertial structure to be localized is that of Newton and Galileo, or Einstein and Minkowski). The concrete theories of general relativity and of Newtonian gravity are then described within those frameworks.

(Other general frameworks cross-classify these frameworks of spacetime and inertia. *Particle mechanics* is the framework in which lie all theories that describe systems of point particles interacting with one another, whether those 'point particles' are genuinely tiny or are idealizations for stars and planets; it may be contrasted with *field theory*, the general framework of which the theory of electricity and magnetism is a special case.)

Yet even frameworks as general as these are not the limit. *Classical mechanics*, understood in its broadest terms, encompasses essentially every theory we have so far considered: it makes no assumptions about inertial structure or about the nature of matter. In essence, it requires of theories only (i) that each system they describe has a physical state—a description of what the system consists of and what properties it has—whose evolution over time is the business of the theory to state; and (ii) that when a system can be broken into parts, each part has its own state, so that the state of the whole can be given by giving the states of the parts.

Is that really a framework, you may ask, or a mere tautology? Is it even conceivable that a physical theory might *not* fall under the classical-mechanics framework? Remarkably, it is not just conceivable but actually true: a vast part of physics, including most of the great successful theories of the 20th century, lies instead in the framework of *quantum mechanics*. Originally developed to account for sub-atomic physics, it now underlies physics at every scale from the Higgs boson to the quantum fluctuations that distributed galaxies across the night sky, pausing on the way to encompass, to name three, the theory of superconductivity, the workings of the transistors inside every laptop and mobile phone, and the secrets of nuclear weapons. It is, to put it mildly, a pretty successful framework.

And yet, we do not understand it.

Or rather: there is no *agreement* as to how to understand it, beyond a consensus that at any rate it cannot be made sense of in the way that classical mechanics can. Some physicists and philosophers treat quantum mechanics as a reason to rethink the whole conception of science. Others think its paradoxes are so severe that the theory itself must be replaced. Still others see in it evidence that our own Universe is only one part of a far vaster reality.

In this chapter, through descriptions of three simple experiments, I will try to convey just what these deep issues in quantum mechanics really are, and why the quantum framework is so different from the classical. In Chapter 6, I will see what lessons we might learn from this for physics, for philosophy, and for our understanding of our place in the cosmos.

Interference and measurement

Shine a laser at a light detector, and measure the strength of the signal as the power of the beam is reduced. For a while, the result is simple: just a smooth decrease in the amount of laser light detected. But there is a threshold energy where, if light is not detected at that energy, it is not detected at all. Turning the laser power down, beyond that point, doesn't mean weaker and weaker detection: it means that detection happens less and less often, but each time it does happen that same, fixed, amount of energy is detected. Light, that is, seems to come in chunks of fixed energy—*quanta* of light (hence, *quantum mechanics*). The energy of each light quantum depends only on the colour of the laser: for a given colour, more or less energy corresponds to more or fewer quanta being emitted. Or put another way: a beam of light is a stream of light *particles*—or *photons*, as physicists call them.

So far so good: there is nothing intrinsically crazy about light being made of particles (it's what Newton thought, in fact). Now we make things slightly more complicated. A *half-silvered mirror*—a mirror that reflects half the light that falls onto it—is placed at an angle into the laser beam, splitting it into two half-strength beams (Figure 6(a)).

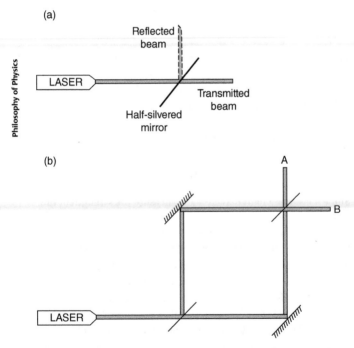

6. Interference experiments with light: (a) splitting a light beam with a half-silvered mirror; (b) interference—splitting and recombining a light beam.

Does the half-silvered mirror split each photon into two half-strength photons? Or do half the photons go one way, half the other way? At first sight, either is possible, but when we look—when we put detectors into each beam—we find that every time a photon is detected, it has the same energy—depending only on the colour of the beam—that we found earlier. So (it seems) each time a photon hits the half-silvered mirror, it either gets deflected or let through, with a 50 per cent chance of each. Again, so far so good.

One more complication. We use ordinary mirrors to bend the two beams around so that they cross, and at their crossing point we insert another half-silvered mirror (Figure 6(b)). What this does is: it splits the beams, then recombines them, and splits them again. And it now looks as if there are *four* ways each photon can travel. If it bounces off *both* half-silvered mirrors, or off *neither*, it ends up in one place (A in Figure 6(b)). If it bounces off *the first half-silvered mirror but not the second* or off *the second but not the first*, it ends up in another (B in Figure 6(b)).

We know that a photon has a 50 per cent chance of bouncing each time it hits the mirror. So it looks as if each of these four possibilities is 25 per cent likely. Even without doing the experiment (it seems) we know what we must find: half the time (25% + 25%) the photon gets to A and half the time to B, so the two detectors give equal-strength signals.

Yet this is not what we find. Depending very delicately on exactly how long each beam is, the experimenter can arrange for *all* the light to be detected at A, or *all* the light to be detected at B, or anything in between.

Physicists call this effect *interference*, and they mean it literally—one of the beams somehow interferes with the other, so that what happens to the light depends on *both* beams. And indeed, if we block one beam entirely, the interference goes away—half the (remaining) light turns up at A, half at B. So this seems to suggest that what's really going on is an interaction between the photons of light, so that the ones in the left-hand beam bounce off the ones in the right-hand beam.

That too is testable. We can turn down the strength of the laser (or, more realistically, put a nearly opaque filter in front of the laser) until only one photon is going through at a time. If interference happens because some photons bounce off other photons, this ought to mean that the interference effect goes away when only one photon at a time is present. But it doesn't: however weak the beam, the interference effect persists unchanged. If we have arranged for all the photons to be detected at A, that is where they will be detected whether one passes through at a time, or a trillion do.

So what is going on? It looks as if *something* must be in both beams *whenever* a photon passes through the system. The 'something' behaves just like a photon: doing anything to the beam that blocks photons, stops the interference; doing anything to the beam that leaves photons unaffected, leaves the interference unaffected. Yet whenever we look at both beams, we find only one photon at a time, and we find it in one beam or the other—never both. In other words, it seems as if the photon is in *both beams at once*—until we look at it, at which point it makes its mind up to be in one beam or the other.

But this makes little sense. 'Looking', after all, is just one more physical process. The devices we use to measure the presence of photons are themselves just made of microscopic particles, governed by quantum mechanics. And the act of looking is just one more interaction between physical systems, again governed by quantum mechanics. We shouldn't have separate rules for how atoms behave according to whether we humans designate a bunch of atoms as 'a photon detector' or not.

To see what quantum mechanics itself says about how the detectors ought to behave, we can resort to an old (and less than ethical) thought experiment due to Erwin Schrödinger. Suppose we simply look to see which beam the photon is in, by putting a detector into each beam. And suppose we wire up the left-hand detector—but not the right-hand one—to a gadget that kills cats, and we put some unfortunate feline in range of the gadget.

If the photon is in the left-hand beam, the cat will die. If the photon is in the right-hand beam, the cat will live. So if the photon is in both beams at once, what quantum mechanics appears to be telling us is that the cat both lives and dies—that it is, crazily, alive and dead *at the same time.*

Of course, experimenters doing this experiment do not report 'I saw a semi-live, semi-dead cat!'; they report either live cats, or dead—just as photons are not seen at both detectors at the same time, but are seen at one detector some fraction of the time, at the other detector the rest of the time. (I hasten to add that, to my knowledge, no-one has literally performed the experiment with real cats!)

Here's a different way of putting the paradox. We started with the idea that light comes in localized particles, so that the state of the laser beam is just given by saying how many photons there are and where they are. When a photon passes through the half-silvered mirror, on this way of thinking, it ends up either in the transmitted beam or the reflected one—we don't know which one, but it's definitely in one or the other, and we can use the language of probability to quantify our lack of knowledge. (Or, if you prefer, to quantify the fraction of the times that the photon is in one beam or another—the puzzles about understanding probability, which we encountered in Chapter 4, occur here too!) We saw that this interpretation founders on the rock of interference, which pushes us towards regarding the photon as being in both beams at once. But then that delocalized, extended conception of the photon itself runs into trouble when we actually measure where it is—at that point, the interpretation of the photon as having an unknown, definitely localized state takes over again, and the interpretation where it has a known, indefinitely localized state has to be dropped. We seem to have to shift, inconsistently, from one description to another and back, depending on what features of the experiment we wish to describe.

The mathematics of interference is formally equivalent to the mathematics of waves, and so this interpretative inconsistency used to be called 'wave-particle duality' (and, indeed, that name persists in older textbooks and in much popular science). But it's better to understand it as an inconsistency in the account of the system that quantum mechanics gives us: we have a choice between (i) a story of indefinite, delocalized, but known properties (the 'indefinite description'); and (ii) a story of definite, localized, but unknown properties (the 'probability description'). Interference seems to require the indefinite description; measurement seems to require the probability description. And the question of how to make sense of the theory, given the apparent need for both (i) and (ii) and the apparent inconsistency between the two, is called the 'measurement problem'.

Superpositions and quantum states

Physics has a language to describe the apparent indefiniteness of photons (or anything else). If a particle can be here or there, it can also be in a *superposition* of here and there. Using the notation developed by the physicist Paul Dirac, we can write this superposition as

$$| \text{STATE} > = a \,| \text{ here} > + b \,| \text{ there} >$$

|STATE> is called the 'quantum state' of the particle, and quantum theory, and its philosophy, are essentially about how

these states evolve and are interpreted. The terms 'a' and 'b' are *amplitudes*, mathematical objects (complex numbers, to be specific) that have both a *magnitude* (a positive, real number, written $|a|$) and a *phase* (which can be thought of as an angle). Effectively, a complex number is a little arrow: its length is its magnitude; its angle to the x-axis is its phase. These amplitudes describe both the probability and the interference aspects of quantum mechanics. As for the probabilities: if we measure where this particle is, the probability of finding it 'here' is equal to the squared magnitude of the amplitude for 'here': Prob(here) = $|a|^2$. And similarly for 'there'. This probability rule is called the Born rule, after Max Born who originally proposed it; pretty much all evidence for quantum mechanics relies on it.

The Born rule makes |STATE> look a bit like a probability distribution: saying that the particle has |STATE> as its quantum state bears a certain resemblance to saying

$$\text{'here'} \left(\text{Probability } |a|^2 \right) \qquad \text{'there'} \left(\text{Probability } |b|^2 \right)$$

But crucially, two quantum states can assign to an outcome amplitudes with the same magnitude but different phases—and the phases affect how the system evolves over time, including how different possible paths the system takes might interfere with one another. Replacing b with –b in |STATE> makes no difference to the probabilities, but can make a big difference to how |STATE> evolves over time—which, in turn, can make a big difference to what the probabilities of future measurements will be.

We see in |STATE> a direct realization of the two conflicting descriptions of quantum systems which we discussed in the last section. When we want to understand the effect of measuring a system, the probability interpretation of |STATE> is natural: if |STATE> is just a mathematical way of saying 'the system might be here, or it might be there; here are the probabilities' then 'superpositions' are unmysterious. But we cannot understand

|STATE> this way if we are concerned instead with a system's dynamics, at least not if those dynamics contain interference: interpreting |STATE> just as a probability distribution amounts to losing track of the phase information, and that information has empirically significant consequences.

But if the probability interpretation of |STATE> looks problematic, the interference interpretation is not much better. Consider again Schrödinger's poor cat, whose state might be written as

$$| \text{CAT STATE} > = (1/\sqrt{2}) \left(|\text{ALIVE} > + |\text{DEAD} >\right)$$

This is an equal superposition of 'alive cat' and 'dead cat', and applying the Born probability rule tells us that a measurement is equally likely to find the cat alive as dead. (The $(1/\sqrt{2})$ term just makes sure that the squares of the amplitudes add up to 1.) On the probability interpretation of |CAT STATE>, none of this is especially mysterious: |CAT STATE> just represents a cat that is as likely to be alive as dead; measurement just tells us which it is.

But if |CAT STATE> represents the actual, physical state of the cat, something seems terribly wrong: that state is a bizarre, half-dead, half-alive cat, of a kind that no-one has ever seen—and the Born rule, which predicts that what we actually find when we measure the cat is a live cat with probability 50 per cent and a dead cat with probability 50 per cent, is nowhere to be seen. (Indeed, probabilities in general are nowhere to be seen if we understand |CAT STATE> this way.)

So: the quantum state apparently cannot be understood *either* as describing the actual physical properties of the system (as in classical mechanics) or as describing the probabilities for the system to have various properties (as in statistical mechanics). The dilemma becomes sharper still when we consider the

Philosophy of Physics

quantum state of multiple systems, and the remarkable property of *entanglement*.

Correlation and entanglement

Quantum particles—electrons, say—have a property called 'spin'. *Very* roughly, spin encodes the rotation of the electron around its axis, but it is stranger than that: its oddities could make up a chapter of this book themselves but I will largely put them aside. Suffice it to say that for any direction in space, the spin of an electron can be measured in that direction, and there are only two outcomes, which we can call 'up' and 'down'. A generic quantum state of an electron (ignoring those features which tell us where it is in space, and focusing only on spin) could then be written

$$| \text{ELECTRON} > = a \,| \text{UP} > + b \,| \text{DOWN} >$$

where |UP> and |DOWN> are relativized to some, fixed, direction in which we choose to measure the spin.

Now let's consider *two* such electrons. If we measure their spins separately, there are four possible outcomes: 'up' for both; 'down' for both; 'up'/'down'; 'down'/'up'. So a generic quantum state of the two electrons together can be written

$$| \text{TWO ELECTRONS} > = a \,| \text{UP,UP} > + b \,| \text{UP,DOWN} > + c \,| \text{DOWN,UP} > + d \,| \text{DOWN,DOWN} >.$$

It's natural to ask: what are the states of the two electrons separately, given this state for both electrons together? Remarkably, there may be no such states. For consider: if each electron has its own state, that state determines (by the Born probability rule) what the probabilities are of getting 'up' or 'down' when spin is measured. And since those probabilities are determined by the state alone, there can be no prospect of correlation between the measurements, no prospect that measuring the spin of one particle gives information about the result of measuring the other.

But it is easy to write down two-electron quantum states for which this is not the case: for example,

$$| SINGLET > = (1/\sqrt{2})\big(| UP,DOWN > - |DOWN,UP >\big)$$

If we measure the spin of the two electrons when their state is |SINGLET>, we have—according, as always, to the Born rule—a 50 per cent chance of getting 'up'/'down', a 50 per cent chance of getting 'down'/'up', and no chance at all that both particles are measured to have the same spin. So there is a perfect anti-correlation between the results of the two spin measurements, and hence no prospect of assigning separate quantum states to the two electrons. States like this are called *non-separable*, or—more poetically—*entangled*: they can only be described jointly, not in terms of the separate features of the system's constituents.

If a pair of electrons has |SINGLET> as its state, and the spin of one of the electrons is measured, the result of measuring the other spin is predictable with 100 per cent success, no matter how far apart they are. Whether this seems bizarre or mundane depends very much on how we think of quantum states. If they are thought of as probabilities, this tight correlation is unmysterious in itself: if you know, say, that a white card and a black card have been put in two envelopes which are then shuffled, and on opening the first envelope you find a white card, you know for sure that the other envelope contains a black card even if it is miles away—but no mysterious non-locality makes it so. If on the other hand a singlet state describes a state of affairs that is simultaneously (up here, down there) and (down here, up there)—whatever that means—then before the measurement the states of the two electrons are indefinite, and after the measurement they are definite—and that chance from indefiniteness to definiteness seems to occur instantaneously, no matter how far apart the electrons are.

This might seem to support the probability interpretation of the state, and to support the idea that 'entanglement' is just another word for correlation. But—even if we set aside the profound difficulties that interference poses for a probability interpretation of quantum states—entanglement cannot be understood so simply, as we will now see through consideration of a simple game and its remarkable implications.

Bell's theorem and the necessity of non-locality

Cooperative games—games where the players work together towards a shared goal—are in fashion at the moment. The 'Bell game', a cooperative game for two players, hardly competes with the most enjoyable of these games, but what it lacks in gameplay it makes up for in philosophical significance. To play, we need two coins, two cards, and two rooms, one of each for each player. The coins can land 'heads' or 'tails' when tossed; the cards have one white side and one black side; the rooms are distant from one another and locked.

Here's how you and I can play a round of the game. We go into the two rooms, and each flips our coin. Having done so, we put our card either white-side up or black-side up. The winning conditions are a little strange: we win if we both put the card the same side up (both white, or both black)—*except* that if we both get 'tails' when we flip the coin, we win if our cards are different sides up (black/white or white/black). We play many rounds, and then at the end score up: our score is the fraction of times we won. Here's a summary of the rules:

My coin	Your coin	Winning condition
Heads	Heads	Both cards same way up
Heads	Tails	Both cards same way up
Tails	Heads	Both cards same way up
Tails	Tails	Both cards opposite ways up

Before starting to play, we can compare notes, so that we can come up with the best strategy we can. A pretty good strategy, for instance, is just to agree that we will always put the card white-side up, no matter what. On that strategy, we'll win three-quarters of the time, losing only when the two coins are both tails.

What if we want to do better—to win not just three-quarters of the time but all of the time? It isn't hard to convince yourself that it's impossible. Suppose, for instance, that we agree that you will play white, whatever happens. I need to play black if both of our coins are 'tails', but white if only one of them is—but to work that out, I need to look at your coin as well as mine, and it's in a different room. So I can't avoid sometimes playing the wrong card, so that at least one of the four possible combination of coin tosses leads to us losing. Similar problems occur whatever strategy we try: because we can only look at our own coin, and because the winning conditions depend on a joint property of both coins, there will always be at least one of the four coin-toss outcomes on which we lose. We can't get above 75 per cent.

What about if we allowed randomness in our strategies? We might decide, in advance, to play a different strategy on each round of the game; we might even make our plan by tossing coins or rolling dice of our own. It won't help. None of the 'pure' (that is, non-random) strategies do better than 75 per cent success, so randomly mixing those strategies can't get us above 75 per cent either.

So suppose that we come across two players of this game who *do* manage to beat 75 per cent. How could they be doing it? The obvious possibility is that they are cheating—they have sneaked a pair of mobile phones or similar into the rooms and are comparing notes. We have proven, it seems, that unless you cheat in this way, the best score possible in the Bell game is 75 per cent—so a score above 75 per cent is proof of cheating.

Suppose we *really* care about preventing cheating. Here's an apparently sure-fire way to do it: make sure that the two rooms are so far apart that even light doesn't have time to travel from one room to the other during the course of the game. (Maybe one room is on Earth, one is in orbit around Jupiter, and we only play for ten minutes, less than the 35–50 minutes it takes light to travel between the planets.) At this point, cheating seems physically impossible: it would require faster-than-light signalling.

Our result—that 75 per cent is the maximum possible score on the Bell game without being able to signal between the two rooms—is called *Bell's inequality*, after the physicist John Bell. If that score could be beaten, it would seem to mean that some signal is travelling between the two rooms in which the game is played; if those rooms are too far apart for light to travel between them, that signal must be a faster-than-light signal.

In a remarkable experiment in Paris in 1986, Alain Aspect set up apparatus that, in effect, played Bell's game and won—not all the time, but more than 75 per cent of the time. (He used randomizer devices rather than human players, and the detailed form of the inequality he violated looks different from the 75 per cent we have used here, but the underlying ideas are the same.) Since then, Aspect's result has been replicated time and again. Here is how the trick is done: we generate pairs of particles in the |SINGLET> state, and send them into the two rooms. In each room, the player (or rather: the automated circuitry acting as the player) measured the spin of the particle in one of two possible directions, with the choice of direction determined at random by the coin flip (actually, by a mechanical randomizer device). A result of spin up is interpreted as playing 'white', a result of spin down, as 'black'. Unlike in our earlier discussion of |SINGLET>, these directions of measurement are not the same for each particle, so we would not predict perfect anti-correlation of the results: what quantum theory predicts is that the anti-correlation gets gradually weaker as the directions of measurement diverge.

The relevant mathematics is a bit beyond the level of this book, but the end result is easy to describe: if we carry out this protocol, preparing and measuring sequences of spins, the score on the Bell game is about 85 per cent—well above the Bell-inequality threshold.

The first thing we can conclude from this is that there is more to entanglement than probabilistic correlation. If |SINGLET> could somehow be thought of as describing a pair of anti-correlated but definite spins, measurements of those spins would just constitute a (very complicated) randomized mixture of strategies—and we have seen that no such mixture can crack the 75 per cent threshold. Whatever entanglement is, it is stranger than that, and seems to have something genuinely non-local about it. The correlations between spin measurements when we measure |SINGLET> are too strong to be attributed to any underlying local description.

But we can learn far more than this from the experiments that violate Bell's inequality. After all, our derivation of the inequality made no use of quantum mechanics: it was simply a demonstration that any strategy that beats the 75 per cent score on the Bell game must be using some kind of faster-than-light interaction. And Aspect's experiment, and its successors, *did* beat that score. So—even if quantum mechanics were to be disproved tomorrow—that experiment seems to be direct empirical evidence that *the world* includes processes that happen faster than light, indeed processes that happen arbitrarily fast, instantaneously even.

That conclusion is hotly contested. Most physicists do not accept that faster-than-light interactions occur in nature: they point both to the tension between special relativity and the existence of such interactions, and to the 'no-signalling theorem'—a straightforward result of quantum mechanics—which demonstrates that at any rate no physical process consistent with quantum mechanics can be used to send actual, usable information faster than light. (So if there are faster-than-light effects underlying the violation of Bell's inequality, they are hidden away, almost conspiratorially, so that we cannot detect them directly). But how the impossibility of faster-than-light signalling can be made compatible with Bell's inequality and its violation is controversial and unclear.

These mysteries—the problem of measurement and of the interpretation of the quantum state; the nature of entanglement; the non-locality implied by violation of Bell's inequality—are the 'facts on the ground' that any attempt to make sense of quantum mechanics must address. It should already be apparent that a simple extension of the classical-mechanics (or indeed the statistical-mechanics) framework is ruled out: to understand quantum mechanics apparently requires a shift in our philosophical attitude, in the physics itself, or in both. In Chapter 6 we will see how this might be done, and why it matters.

Chapter 6
Interpreting the quantum

Chapter 5 was in a sense mostly negative: I tried to make clear, and vivid, just how strange quantum mechanics is and how severe the obstacles are in the way to understanding it. But that does not mean that understanding is impossible. Ever since the birth of quantum mechanics, physicists and philosophers have been discussing its meaning, and the last forty years have seen really substantial progress in understanding the options—even if that progress has not been matched by the development of any consensus.

In this final chapter, I will lay out and discuss some of the most interesting and popular strategies that have been developed to make sense of quantum mechanics. I should admit that in my view the last strategy I discuss, the Everett interpretation, is most likely to be correct. But my goal in this chapter is less to defend one particular approach as to show how this question is of philosophical *and scientific* interest. Working out how to think about and make sense of quantum mechanics is important: it has led to very substantial scientific results and is likely to lead to more.

Probabilities and measurements

Recall my description of how physics in practice treats the quantum state: inconsistently, as either a probabilistic description of a system's unknown but definite state, or as a physical description of its indefinite state. Most ways to make sense of quantum mechanics can be thought of as committing to one approach or other, and trying to resolve the apparent paradoxes of that approach. Here, we'll begin with the probabilistic approach.

Let's consider again the spinning electron, as a paradigm quantum system: a generic state of that electron can be written as

$$|STATE> = a|UP> + b|DOWN>$$

and the Born probability rule tells us that if we measure its spin (as usual, along some fixed axis), we have probability $|a|^2$ of getting 'up' and probability $|b|^2$ of getting 'down'. As I explained in Chapter 5, this probability rule cares only about the magnitudes of a and b, and not about their phases; those phases matter because they influence how the state evolves, and in particular how interference plays out.

However, there is a way of getting at those phases through measurement, as long as we remember that more than one thing can be measured. Suppose we instead measure spin along a new axis, at right angles to the old one (let's say the old axis is the z-axis and the new one is the x-axis). Then the laws of quantum theory say that this same state can be written as

$$|STATE> = (a + b) /\sqrt{2}|UP;x> + (a - b) /\sqrt{2}|DOWN;x>$$

so that the probability of getting 'up' on an x-axis measurement is $(|a + b|^2 / 2)$, which depends not only on the magnitudes of a and b but also on their phases. (Here $|UP;x>$ and $|DOWN;x>$ are states with spin up and down about the new axis.) So the phases need not just be thought of as carrying dynamical information; they also carry information about the results of other measurements.

111

(Actually, these are just two sides of the same coin. One way to measure spin along the x-axis is just to rotate the system by 90 degrees, so that the x-axis turns into the z-axis, and then measure spin along the z-axis. The ability to make measurements about any axis is equivalent to the ability to apply arbitrary dynamical transformations to a system and then make measurements about a fixed axis.)

All of this generalizes to other directions of measurement, and indeed to other quantum systems. It is not difficult to show that, if we are given the probabilities for each outcome of *any* measurement that could be performed on a system, then that is enough to work out the complete quantum state.

What more would be required to make sense of the quantum state in probabilistic terms? Just this: some assignment of actual properties to the system such that (a) quantum measurements can be understood as passive reports of what those properties are, and (b) the quantum state can be understood as determining the probability that the system has a given collection of properties. (This is what classical-statistical mechanics gives us: the underlying properties are the actual positions and velocities of the particles that make up the system; statistical probabilities encode how likely it is that those positions and velocities have any given value.)

We saw in Chapter 5 that interference seemed to prohibit doing this in any straightforward sense: the photon couldn't be in one channel or another on pain of failing to account for interference, and couldn't be spread across both channels on pain of failing to explain why it is always measured to be in one channel or another. In fact that argument can be sharpened and made much more precise. Powerful mathematical results—the Kochen–Specker theorem, Gleason's theorem, the Pusey–Barrett–Rudolph theorem—have by now convinced (nearly) everyone in the field that no such strategy is possible. (And the Bell inequality tells us that any such strategy would in any case involve faster-than-light interactions.)

But there is an alternative: hold on to the idea that the quantum state is understood in terms of the probabilities of various measurement outcomes, but abandon the idea that those measurement outcomes are reports on the pre-existing properties that the system has. From this perspective, the quantum state is a mathematical device used to summarize what happens when physicists perform various processes in the lab; any attempt to understand those processes as measurements of some underlying reality—or indeed to understand quantum theory as a description of the world in itself and not just as an algorithm to predict measurement outcomes—is set aside.

This approach to understanding quantum mechanics is a variant of *instrumentalism*, one of the philosophies of science which we considered in Chapter 1: we are to think of quantum mechanics not as a description of the world, but as a tool to describe the results of experiments. Questions about what the system is doing when we don't measure it—for instance, what Schrödinger's poor cat is up to before we open its box—are set aside, on instrumentalist approaches, as meaningless: we ask them only if we have not understood what sort of theory quantum mechanics is.

Proposals like this have been around in quantum theory since the 1920s (Niels Bohr and Werner Heisenberg, two of the founders of the subject, were to varying degrees sympathetic); they remain popular in some parts of the physics community. The great majority of philosophers are sceptical, given the problems with instrumentalism we considered in Chapter 1: it relies on a separation between the 'observational' part of a theory (about which the theory actually makes meaningful claims) and the 'theoretical part' (which is just a tool to help us analyse the observational part), and that separation does not match physics as we find it.

In the particular context of quantum mechanics, the problem is this: quantum measurement devices are not black boxes, scattered across the desert by benevolent aliens or deities. They are complicated physical devices, built to interact in complicated ways and relying, themselves, on the principles of quantum mechanics. We cannot understand what a measurement device is or what it is measuring, or even if it is measuring anything at all, unless we understand its workings—in which case we need a way of understanding quantum mechanics in order to do so, and on pain of circularity that 'way of understanding' had better not presuppose we know what the measurement devices are.

Furthermore (though really it is a variant of the same objection) many of our applications of quantum theory do not fit well to the context of a lab and a measurement of a state. Many of the triumphs of quantum theory concern its explanations of the macroscopic properties of matter—why metals conduct, why gold is the colour it is, how crystals behave when heated—and those explanations can't easily be analysed into discrete measurement predictions. For an even more dramatic example, consider the quantum fluctuations in the early Universe that gave rise to the distribution of matter over the largest scales—the theories of those fluctuations are testable, but only as part of a vast and complex theoretical framework for cosmology, so that there is no remotely simple way to argue that observations of the distribution of galaxies are just a 'measurement' of the quantum state of the early Universe.

These objections are far from conclusive—and given that less than a century ago instrumentalism was the dominant philosophy of science, it would be hubristic for philosophers to be fully confident that instrumentalist approaches to quantum mechanics should be ruled out. But they at least give us strong reasons to investigate the alternative approach: accepting the quantum state as a description of a system's physical properties, and reconciling that with the problem of measurement and the paradox of Schrödinger's cat.

Change the physics?

There is a (conceptually) very straightforward alternative way to approach the paradoxes of the quantum: we could decide that they are not just paradoxes but *contradictions*, demonstrations that quantum theory is *wrong*. Any theory that predicts that cats are alive and dead at the same time, when manifestly they are not, might be said to have refuted itself: perhaps the issue is not how to understand quantum mechanics but how to modify it—or replace it—so that it is not in flat contradiction with the facts. Given how fantastically successful the theory is, of course, any such modifications will have to be done delicately, to preserve those successes—and this is easier said than done.

There are a vast number of proposed strategies for how quantum mechanics can be modified, but I will focus here on just two prominent examples—*dynamical collapse* and *hidden variables*. The starting point for the first is the Schrödinger cat state, which (generalizing a bit to allow for variable amplitudes) can be written as

$$|CAT\ STATE> = a|ALIVE> + b|DEAD>$$

Since (it is argued) this state is not what we find when we actually look at the cat, the theory needs to be modified so that states like this do not arise, or at any rate do not persist when observed. This amounts to changing the equations of quantum mechanics, to introduce a new evolution that can be written as

$$|CAT\ STATE> \rightarrow |ALIVE> \text{ (with probability } |a|^2)$$
$$|CAT\ STATE> \rightarrow |DEAD> \text{ (with probability } |b|^2)$$

If this transition has always occurred by the time that we actually observe the cat, it resolves the measurement problem—we find the cat either alive or dead (and not in a weird superposition of both), and the probability of each matches what the Born rule predicts. We can call this process *quantum state collapse* (other terms are 'wavefunction collapse' and 'state vector collapse', named for various mathematical ways of thinking about the quantum state).

Quantum state collapse was introduced in the infancy of quantum theory—Paul Dirac, one of the founders of the theory, proposed that it should occur *exactly* when a system is measured. That is, there should be two different sorts of quantum-mechanical dynamics—the ordinary sort (which physicists call 'unitary'), which applies whenever a system is not being measured, and the collapse rule, which occurs at the point of measurement. If we adopted the probabilistic reading of the quantum state (where the Schrödinger cat state just represents the probability of the cat being alive or dead), it would be trivial, corresponding only to our update of information when we actually discovered whether the cat survived. But as a proposed modification of quantum theory to solve the measurement problem, it cannot be thought of that way: instead, it is an instant, random change of the actual state the cat is in, shifting it from its undead indefiniteness to a more reputable state, either alive or dead but in any case definite. (As such, if we introduce quantum state collapse to solve the measurement problem, we do so as part of a *physical* interpretation of the quantum state—the probabilities now occur because of the randomness in the collapse rule, not as part of the very interpretation of the state.)

This way of presenting quantum mechanics is still found in introductory textbooks—but it has largely been abandoned in the actual practice of quantum mechanics. Its basic problem is that it treats 'measurement' as a primitive, unanalysed notion, and we saw in the last section that this conflicts with physicists' desire to treat measurements as physical processes open to study with the tools of quantum theory itself. Another way to put this, in light of the discussion in Chapter 4, is that a theory which has 'measurement induces quantum state collapse' as one of its fundamental principles is a theory in which any reductionist analysis of measurement in simpler terms is ruled out—and yet physicists seem to make such analyses all the time, and indeed to rely on them in building the complicated devices actually used in physics labs.

But there is an alternative way to think of quantum state collapse: instead of including a fundamental posit that collapse occurs upon measurement, we could imagine a theory where collapse occurs for some other reason, in some other circumstances, that can be described and defined sharply in terms of microscopic physics—and yet those circumstances are *in fact* such as to ensure that collapse has occurred well before actual measurements are completed. Theories of this kind are known as *dynamical collapse theories*—'dynamical' in reference to some microscopically defined, bona fide dynamical mechanism, instead of collapse by definition being triggered by the macroscopic concept of 'measurement'.

The constraints on any such theory are strict. If collapse happens too soon, it will suppress the interference effects which quantum theory relies on for its predictions and explanations, and so will falsify itself. If it happens too late, it will fail in its duty to suppress Schrödinger cat states. But, at least for simple versions of quantum mechanics, some such theories have been developed. They make predictions that deviate from those of 'normal' quantum mechanics (they predict that interference fails to occur in certain exotic, but in-principle-testable, circumstances)—so far experiments have failed to find any such deviations, but the experiments are extremely difficult to perform and it is not possible to rule out progress in the future.

The second strategy for modifying quantum theory starts with the apparently dual nature of the quantum state, both physical and probabilistic. It makes sense of this by supplementing quantum theory with additional 'hidden variables', whose role in the theory is to describe the actual measurement outcomes: in the Schrödinger cat state, for instance, the quantum state remains indefinite but it no longer has the task of representing the macroscopic, observed world. That task falls to the hidden variables, which represent either a live cat or a dead cat—until we know what the values of those variables are, though, the cat might be either alive or dead. In these *hidden-variable theories*, probability enters the theory in the same (controversial!) way that it does in statistical mechanics, and so the probabilistic aspects of quantum theory are moved from the quantum state to the hidden variables.

Some hidden-variable theories have been constructed, again at least for simple versions of quantum mechanics (the most famous is the *de Broglie–Bohm theory*, sometimes called *Pilot-Wave theory*, or *Bohmian mechanics*, named for physicists Louis de Broglie and David Bohm). They are fairly popular among philosophers; those physicists inclined to modify quantum mechanics have more often tended to favour dynamical-collapse theories, I suspect because hidden-variable theories normally do *not* make predictions that conflict with orthodox quantum mechanics, and so (to most physicists' eyes) the hidden variables complicate the theory without any empirical payoff. Advocates would retort that the payoff comes in having an intelligible theory in the first place.

(There is a more purely philosophical problem with hidden-variable theories. They rely—at least as they are usually presented—on the assumption that macroscopic measurements actually detect the hidden variables, and not properties of the quantum state. (For instance, it is critical that measurements of the cat detect the hidden variables, which represent either 'live cat' or 'dead cat', and not the quantum state, which if it represents anything intelligible represents both a live cat and a dead cat.)

In the most commonly discussed variants of hidden-variable theories, it is a premise of the theory that this is so, not something that can be dynamically derived from any analysis of measurement processes available within the theory. And this seems again to require a primitive notion of how measurement connects to physics, akin to that required by instrumentalism. This is highly contested territory in the philosophy of physics, and turns on fairly deep questions about what it is to interpret a physical theory.)

Hidden-variable and dynamical-collapse theories raise interesting philosophical puzzles, but the main problem with either is the enormous successes of quantum theory. To date, neither class of theories—nor any other approach that relies on modifying quantum theory—has succeeded in reproducing quantum theory's predictions outside a relatively narrow range of applications: roughly, those concerned with the physics of matter moving at non-relativistic speeds, in situations where light can be ignored. No modificatory strategy, for instance, at present can reproduce the two-slit experiment (which uses photons of light), or explain the workings of a laser, let alone make sense of modern particle physics. Some progress has been made on this problem (exactly how much progress is contested; my own view is fairly sceptical), but at any rate, advocates of modifying quantum theory to solve the measurement problem are committed to a truly massive reconstruction project, rebuilding most of 20th-century physics on a new foundation.

Many worlds

The approaches of the previous section might be called 'change-the-physics' approaches: they maintain a baseline scientific realism as the way to make sense of a physical theory, judge quantum theory according to that standard, and find it wanting; the solution is to change quantum theory itself. By contrast, the probability-based approaches to understanding quantum mechanics are mostly 'change-the-philosophy' approaches: they leave the formalism of quantum theory unmodified, but adopt a different (normally somewhat instrumentalist) approach to what it is to understanding a physical theory.

In principle, both change-the-physics and change-the-philosophy are reasonable ideas. The paradoxes of quantum theory tell us that *something* is wrong; both physics and the philosophy of science seem reasonable places to look for that something. Indeed, there might seem to be a natural division of labour: philosophers are best placed to re-evaluate alternatives to scientific realism in the face of quantum paradox; physicists are best place to explore alternative physical theories.

In practice, this is not the way it goes. Very few philosophers (there are exceptions) really take seriously the idea that solving the measurement problem requires us to change our philosophy of science; by contrast, philosophers who think about these issues very commonly conclude that they show a deficiency in physics. Very few physicists (there are exceptions) really take seriously the idea that solving the measurement problem requires us to modify quantum mechanics itself; by contrast, physicists who think about these issues very commonly conclude that they require a new and more imaginative philosophy of science. The obvious interpretation is that philosophers are sensitive to how difficult the change-the-philosophy strategy is but complacent about the difficulties of the change-the-physics strategy—and vice versa.

But since the combination of scientific realism and unmodified quantum mechanics seemed to lead us to the paradox of Schrödinger's cat, it might seem that *either* the philosophy of science, *or* the formalism of quantum mechanics must be modified, however difficult that might be. In fact this is not the case: there is a third option. The basic idea is due to physicist Hugh Everett in 1957, hence its official name: the 'Everett interpretation'. To see how it works, consider Schrödinger's cat yet again, and ask *how* we know that a system can't after all be simultaneously in a live-cat and a dead-cat state. The obvious answer is that we never see cats in such states—but 'seeing' is a physical process, and so we need to model it physically, within quantum mechanics, to determine what actually happens when observers—when I, for instance—interact with Schrödinger's cat.

Here is a simple way of doing so. Very schematically, I must have at least three relevantly distinct states: |IGNORANT> (the state I am in before seeing the cat); |SEES ALIVE> (the state I enter upon seeing a live cat); and |SEES DEAD> (the state I enter upon seeing a dead cat). Suppose I look at a definitely living cat; before the observation, the joint state of cat and me will be |ALIVE; IGNORANT>, and this state will evolve into |ALIVE; SEES ALIVE>:

$$\text{ALIVE; IGNORANT>} \rightarrow \text{|ALIVE; SEES ALIVE>}$$

Similarly, if the cat is definitely dead to start with, the process of observation must proceed like this:

$$\text{|DEAD; IGNORANT>} \rightarrow \text{|DEAD; SEES DEAD>}$$

Now suppose the cat starts in the Schrödinger cat state,

$$\text{|CAT STATE>} = a\text{|ALIVE>} + b\text{|DEAD>}$$

Intuitively, we might expect my observation of this system to look something like

$$\text{|CAT STATE; IGNORANT>} \rightarrow$$
$$\text{|CAT STATE; SEES WEIRD INDEFINITE CAT>}$$

But intuition is a poor guide to physics, and what the physics actually tells us (as an automatic consequence of how observations of definitely living and definitely dead cats go) is that |CAT STATE; IGNORANT> can be rewritten as

$$|CAT\ STATE;\ IGNORANT> =$$
$$a|ALIVE;\ IGNORANT> + b|DEAD;\ IGNORANT>$$

and so it evolves like this:

$$|CAT\ STATE;\ IGNORANT> \rightarrow$$
$$a|ALIVE;\ SEES\ ALIVE> + b|DEAD;\ SEES\ DEAD>$$

According to quantum mechanics, I do not evolve into a state of definitely seeing an indefinite cat; I evolve into an indefinite state of my own, a state which is at once two ordinary, definite measurement outcomes.

And this goes on. If you ask me whether the cat lives, you end up in two states at once: one of hearing me say 'yes', one of hearing me say 'no'. Indeed, the combined state of all of us—you, me, and the cat—is two states at once, but both individual states are mundane: the state where we come across a live cat, or the state where we come across a dead cat. If a third person enquires, or if you post the cat's status on Facebook, more and more systems get drawn into this two-states-at-once condition—get entangled with us, that is.

Indeed, above a certain scale—a scale much smaller than the poor cat—interaction between one system and another is unavoidable even if there is no intentional 'looking'. The gravitational effect of the cat on the air around me or the particles in my body effectively entangles me, and you, and our surroundings, with the cat, whether or not we attempt to know its condition. Try to put anything the size of a cat into a two-things-at-once state, and before long the whole planet, the whole solar system, will be in a two-things-at-once state.

But what are those two things? Each is crushingly ordinary: they are both normal states of the Earth, differing only in whether some unfortunate cat lived or died. Each one develops in time according to the ordinary rules that govern normal states of the Earth. The state of the Earth, that is, consists of two parallel branches: a 'cat lived' branch, and a 'cat died' branch, each evolving over time without reference to the other.

There is a good word for a part of reality that looks like the ordinary Earth and which evolves without reference to other parts of reality: a *world*. Not a world in the sense of an entire self-contained universe, but a world in the sense that Earth or Mars is a world: they are pieces of reality that interact strongly with themselves but are scarcely affected by one another.

Of course, experiments with cats are scarcely the only place where the effects of quantum theory magnify up to human-sized objects. We live in a universe where small changes at the micro-level can, over time, reach the scale of the everyday. An electron in a fluorescent light is both *here* and *there*, a cosmic ray both does, and does not, strike a DNA chain in a cell ... and, soon enough, the light both does and does not flicker; the cell both does and does not mutate. So this splitting into parallel worlds is commonplace, occurring countless times in a second, all across the Earth.

We are led to the conclusion that, if we take quantum mechanics literally and realistically, the world we live in is one of an innumerably greater plurality—an emergent multiverse—all existing in parallel with one another, each one constantly branching from the others. Hence the better-known name for the Everett interpretation of quantum mechanics: the *many-worlds interpretation*.

In some ways, the Everett interpretation is the strangest of all those we have considered. But in another sense, it is by far the most conservative: it requires no modification of the phenomenally successful quantum-mechanical formalism, and no radical reconsideration of the scientific enterprise. Its admittedly astonishing prediction of a branching reality is a consequence of the quantum formalism itself, not some additional postulate layered on top of it.

It is (to put it mildly) controversial whether the Everett interpretation is viable. The most obvious, and the most frequently encountered, objection is sheer incredulity, but really this is less an 'objection', more an expression of astonishment. Serious critics of the Everett interpretation have tended to focus on two more specific concerns. First: does the formalism of quantum theory really imply the existence of parallel worlds, or is that just a verbal gloss of no physical significance? Second: how do we find probabilities in the theory?

The key to the first problem (sometimes called the *preferred basis problem*, for reasons which I lack space to explain) is to notice that the quantum dynamics of big complex systems rapidly conceal the interference effects that underlie the distinctively quantum features of quantum mechanics. When a quantum system has a great many moving parts—a great many *degrees of freedom*, as physicists say—those degrees of freedom generally become entangled with one another, so as to make the interference effects undetectable.

For instance, suppose we try to demonstrate interference not with a photon, but with a bowling ball. To do so we would have to prepare the ball in some indefinite state—a superposition of two different positions, say: schematically,

$$|BALL> = a|HERE> + b|THERE>$$

But air molecules, passing photons, and the like bounce off a bowling ball in the |HERE> state differently from one in the |THERE> state. So |BALL> is unstable: it very rapidly gets entangled with millions of other particles, so that the bowling ball and its environment end up in a state like

> | ENTANGLED BALL> =
> a|HERE; many particles record HERE> +
> b|THERE; many particles record THERE>

For an experiment—or a natural dynamical process—to detect the interference between the two terms in this superposition, the dynamics of that experiment cannot only act on the bowling ball: it has to act on both the ball and the environment, and to manipulate them in exactly the right way so as to display the interference. In practice, this is impossible.

This process of constant entanglement with an environment is called *decoherence*. It is an aspect of the irreversible macroscopic dynamics discussed in Chapter 4, and as such it has philosophical puzzles of its own—but it provides a dynamical explanation of why interference effects can be neglected when a system is sufficiently complicated, and why in that situation the system can robustly be described as two (or more) mutually isolated systems evolving in parallel, rather than as one system comprising interfering parts. When I said, above, that each of the 'cat lived' and 'cat died' branches evolved over time 'without reference to the other', decoherence was in the background: it is the physical process whereby quantum systems develop their emergent, but objective, branching structure.

(Decoherence can also be understood as showing us why *in practice* we can get away with treating the quantum states of big complicated systems as probability distributions over underlying, definite facts, even though interference means that this interpretation cannot be sustained: once decoherence has set in, those interference effects are undetectable and so can be ignored. For this reason, it is not uncommon—at least among physicists—to claim that decoherence *by itself* solves the measurement problem, without any need for parallel universes. But to make this viable, we still need to shift our interpretation of what the quantum state is, from a physical to a probabilistic one, and decoherence alone does not license this. In practice, attempts to use decoherence to resolve the measurement problem usually end up as variants on the Everett interpretation—though this remains hotly contested.)

The second problem—the *probability problem*—is more difficult to solve. After a Schrödinger cat experiment, and after decoherence, the state of the world is something like

$$a|\text{LIVE CAT BRANCH}> + b|\text{DEAD CAT BRANCH}>$$

But to connect the theory to experiment, the squared amplitudes $|a|^2$ and $|b|^2$ must be interpreted as probabilities—and it is not immediately obvious why this is justified. After all, normally probability enters physics either through unknown microscopic conditions or through fundamentally probabilistic laws—but there is no relevant ignorance of the microphysics in the Everett interpretation, and no fundamental probabilities in its dynamics. (And we cannot, except metaphorically, regard the squared amplitudes as describing how many copies of each branch there are.)

Solving the preferred basis problem requires engagement with the detailed maths and physics of decoherence theory, but the probability problem is more purely philosophical. The squared amplitudes have the right formal properties to act as probability (they obey the axioms of the probability calculus; decoherence guarantees that they behave as if they were 'fundamental' probabilities); the question is: are they really probabilities? Many strategies have been proposed to answer this question, of which the most developed (by the physicist David Deutsch, and by philosophers like Hilary Greaves, Wayne Myrvold, and myself) try to ask what the scientific method would be like for scientists who took the Everett interpretation seriously, and to recover the result that those scientists would treat the squared amplitudes as if they were probabilities. It is a matter of continuing controversy whether these strategies succeed.

But there is a more basic point to make here. Probability is mysterious in physics even outside the context of the Everett interpretation. We have already seen how difficult it is to make sense of the probabilities of statistical mechanics. The supposedly 'fundamental' probabilities of probabilistic dynamics are no less mysterious. We know how to *use* the concept of probability (roughly: test probabilistic theories by measuring relative frequencies; choose actions that make desired outcomes more probable), but beyond that, there is no agreed-upon *interpretation* of probability. We know that it must make sense somehow, because of the role it plays in our science—but if the Everett interpretation is correct, that role is played not by fundamental probabilities but by the squared amplitudes of decohered branches, and has been all along. We need to avoid a double standard here: if physical probability is mysterious in general, that it is mysterious in one particular theory is not an argument against that theory in particular. In this case (as in many others) the strange setting of the Everett interpretation throws into sharp relief already-existing philosophical puzzles.

Non-locality revisited

In Chapter 5, I presented the 'Bell inequality': a constraint on correlations between distant pairs of measurements (or, as I set it up, on the maximum score in a game based on such correlations), under the assumption that those 'distant pairs' were not in direct communication; a constraint, moreover, that is empirically violated, and so seems to imply faster-than-light interactions in any empirically successful physical theory. We can now ask: how does this play out in the approaches to quantum mechanics that we have discussed?

The question gets its clearest answer for dynamical-collapse and hidden-variable theories. These are explicitly committed, in their formalism, to the existence of faster-than-light interactions—indeed, to instantaneous interactions, to 'action at a distance'. In each case, this is a consequence of how the theories' modified or supplemented dynamics apply for entangled particles. Consider again the two particles in a singlet state, and suppose those particles are taken very far away. According to dynamical-collapse theories, measuring the spin of one of the particles will cause the joint state of both particles to collapse, instantaneously affecting the other particle even if it is miles, or light years, away. According to hidden-variable theories, measuring the hidden variable corresponding to one particle's spin will instantaneously influence the hidden variable corresponding to the other particle's spin, again irrespective of how far away it is. (In each case, this instantaneous interaction is unavoidable if we are to reproduce the observed physics of measurements of entangled particles.)

This means that both classes of theory are at least in severe tension with the theory of relativity—and, with the aid of our work in Chapter 3, we can see why. To say that the effect on the distant particle is instantaneous is to say that the measurement has effects simultaneously for both particles—and we have seen that relativity does not permit any absolute, reference-frame-independent concept of simultaneity, and indeed that it strongly implies that 'simultaneity' is a purely conventional notion. Action at a distance seems to conflict with this, and to reopen questions of the structure of spacetime which relativity seemed to have settled. (The matter is not entirely settled, though: there are, admittedly very simplified, models of dynamical-collapse theories which seem to work around the problem and to remain compatible with relativity.)

Advocates of these approaches complain that their critics fail to see the force of the arguments from Bell's inequality. Those arguments, together with the experimental violation of those inequalities, tell us that *any* empirically adequate theory will need to have faster-than-light interactions—and so (they go on) it is a strength, not a weakness, of these approaches that they honestly, explicitly, demonstrate how the faster-than-light interactions are fitted into the physics.

This is a little too quick, though: the Everett interpretation offers a route around the Bell inequality. For there was a suppressed premise in my presentation of that inequality: I assumed that experiments had unique outcomes, which is not the case in the Everett interpretation. It is generally accepted that experimental violations of Bell's inequality do not force non-locality on a many-worlds theory—and indeed, the unmodified dynamics of quantum mechanics, adopted unchanged by the Everett interpretation, involve no action at a distance, and so no conflict with relativity.

What of the probability-based approaches? Here matters are more complex. For 'realist' probability-based approaches—those where the probabilities are probabilities over underlying, unknown properties—it is accepted that those unknown properties must interact amongst themselves faster than light. But some versions of the instrumentalist approach—notably the 'QBism' version developed by physicists Chris Fuchs, Rudiger Schack, and David Mermin—apparently bypass the Bell inequality. They do so at a heavy price: they deny that it even makes sense to talk objectively about the results of multiple measurements made by distant observers. To the QBists, physics is concerned only with the description of a single scientist's observations—and since that 'single scientist' can't be at two places at once, the Bell inequality does not (it is claimed) apply. Most philosophers are sceptical that this makes sense; I confess to sharing that scepticism.

Why does it matter?

Of the approaches I have discussed so far, the instrumentalist approach and the Everett interpretation are roughly equally popular among physicists (as I have noted, change-the-physics approaches are much less so). But by far the most common position in the physics community is what physicist David Mermin calls the 'shut-up-and-calculate interpretation': the view that we should not worry about these issues and should get on with applying quantum mechanics to concrete problems.

In its place, there is much to be said for 'shut up and calculate'. Not everyone needs to be interested in the interpretation of quantum mechanics; insofar as a physicist working on, say, solar neutrinos or superfluidity can apply the quantum formalism without caring about its interpretation, they should go right ahead—just as a biochemist may be able to ignore quantum mechanics entirely, or a behavioural ecologist may be able to ignore biochemistry. Division of labour is unavoidable in science, and often desirable.

But there is a more aggressive reading of 'shut up and calculate'— not just as a description of a physicist's own approach, but as an exhortation to the community to stop wasting their time. That exhortation is often accompanied by the claim that since all the 'interpretations of quantum mechanics' give the same predictions anyway, it is pointless or even unscientific to worry about which one is correct.

There is a high-minded response to such scepticism: quantum theory tells us about the deepest nature of reality; how could we not be interested in its nature? But there are also more practical things to say, for the claim that the question is unscientific is based on a too-simplistic philosophy of science. We saw in Chapter 1 that underdetermination—where two distinct theories make the same predictions—is a subtle matter: rivalries between such theories are often not resolved by a single crucial test but by the development, over time, of those theories as they are adjusted to predict and explain wider classes of phenomena, and we can see this playing out in the debate about interpreting quantum mechanics.

This is most obvious for 'change-the-physics' approaches. These are proposals for, genuinely, mathematically distinct theories. In some cases, these theories already make predictions—albeit difficult to test—that differentiate them from unmodified quantum theory; in others, they are the seeds of research programmes which may lead in a testably different direction from quantum theory. One can regard this as promising or unpromising science, but it is clearly, recognizably, science. (Which is not to say that all advocates of these theories treat them this way—a good test of how serious an advocate of a dynamical-collapse or hidden-variable theory is for their proposal as *science* is whether they welcome or resist the implication that their theory might have testable deviations from quantum mechanics.)

But even within those approaches which leave the formalism unchanged—roughly speaking, Everett-type approaches based on decoherence and the emergence of a classical branching structure, and approaches which treat the quantum state as probabilistic—there are major differences of scientific method. The Everett interpretation has generally been applied in situations where the goal is to understand how systems evolve and develop when left to themselves. It has been central in our understanding of quantum/classical transitions, in environments ranging from the present-day laboratory to the early Universe; it provides a framework and a language to handle situations where 'experiment' and 'measurement' do not have a clear meaning; its language of 'branches' and 'worlds' has been valuable in quantum cosmology and in non-equilibrium statistical mechanics; it treats the framework of quantum theory as a given and uses it to understand and explore issues in specific quantum theories; it is the dominant approach in high-energy physics and in string theory. Probability-based interpretations have been applied more in situations where the goal is to understand the interventions and manipulations we might make on a system; they are extremely well-suited to the study of computability and information processing, where they have inspired much insightful work; it

Interpreting the quantum

naturally leads us to ask why the framework of quantum theory is what it is and not something else; they are the dominant approaches in quantum information theory.

This is not to say that everyone who has used decoherence-based methods to study cosmology is explicitly committed to the Everett interpretation and its language of 'many worlds', or that everyone who was inspired by a probabilistic interpretation of quantum theory to prove a valuable theorem in quantum information is explicitly committed to one or other variety of instrumentalism. It *is* to say that there has been a continuous flow of ideas and inspiration from considerations of the quantum measurement problem to more concrete issues in quantum theory, and back again. Our present-day understanding of quantum mechanics—as with our present-day understanding of any of the deep theories in physics—owes much to those who went ahead and calculated, even where the basis of those calculations was conceptually unclear. But it owes much, too, to those who, while calculating, chose not to shut up, but instead to think clearly about what the physics meant. And there is no reason to think that this process has ended—no reason not to expect that the interface between calculation and philosophy in quantum theory, and in physics more widely, will lead to further deepening of our understanding of these remarkable, but not quite inexplicable, theories.

References

Introduction

The Daniel Dennett quote is from Blackmore, Susan (ed.) (2005), *Conversations on Consciousness*. Oxford University Press (Oxford), p.91.

Chapter 1: The methods and fruits of science

'Try not to say anything false': Fodor, Jerry (2008), *LOT 2: The Language of Thought Revisited*. Oxford University Press (Oxford), p.4.

Chapter 2: Motion and inertia

'Indeed it is a matter of great difficulty': Newton, Isaac (1689), 'Scholium to the Definitions', in *Philosophiae Naturalis Principia Mathematica*, Bk.1, translated by Andrew Motte (1729), revised by Florian Cajori. University of California Press (Berkeley, CA, 1934).

'Shut yourself up with some friend': Galileo (1632), *Dialogue Concerning the Two Chief World Systems*, translated by Stillman Drake. University of California Press (Berkeley, CA, 1967).

Chapter 3: Relativity and its philosophy

'spreading time through space': Brown, Harvey (2005), *Physical Relativity: Space-time Structure from a Dynamical Perspective*. Oxford University Press (Oxford), p.21 *et seq*.

Chapter 4: Reduction and irreversibility

'in the process of turning around': Albert, David (1999), *Time and Chance*. Harvard University Press (Cambridge, MA), p.77.

Chapter 6: Interpreting the quantum

'shut-up-and-calculate interpretation': Mermin, David (2004), 'Could Feynman have said this?', *Physics Today* 57, p.10.

Further reading

In philosophy, as in any academic area where there is genuine controversy, the best way to get an understanding of that controversy is to read multiple sources and to read them critically; I have tried to err in favour of readings that disagree with my own take on the subject. The readings I include are at quite varied levels, and I use a star system to indicate this. Unstarred readings are at about the level of this book; single-starred entries are at about the level of an undergraduate degree; double-starred entries are more advanced.

General philosophy of science (Chapter 1)

D. Deutsch, *The Fabric of Reality* (Viking, 1997). A lively, opinionated, non-technical discussion of falsificationism (and much else).

T. Kuhn, *The Structure of Scientific Revolutions*, 2nd edition (University of Chicago Press, 1970). (*)

I. Lakatos, 'Science and Pseudoscience' and 'Falsification and the Methodology of Scientific Research Programs', in *Philosophical Papers* vol. 1 (Cambridge, 1978). (*)

J. Ladyman, *Understanding Philosophy of Science* (Routledge, 2002). A general introduction to the philosophy of science.

J. Ladyman and D. Ross, 'Scientific Realism, Constructive Empiricism and Structuralism', *Every Thing Must Go: Metaphysics Naturalised* (Oxford University Press, 2007), chapter 2. An up-to-date guide to issues in instrumentalism, realism, and structuralism, and a good route into the wider literature. (**)

B. van Fraassen, *The Scientific Image* (Oxford University Press, 1980). An influential critique of realism, and one of the most important recent attempts to defend an observable/unobservable distinction. (*)

Philosophy of space and time (Chapters 2–3)

J. Barbour, *The End of Time* (Oxford University Press, 1999). An insightful and accessible account of the substantivalist/relationist debate, very much from the relationist's point of view.

J. Barbour, *The Discovery of Dynamics* (Oxford University Press, 2001). An extended history of space, time, and motion in physics. (*)

J. Bell, 'How to Teach Special Relativity', *Speakable and Unspeakable in Quantum Mechanics*, 2nd edition (Cambridge University Press, 2004). Classic, though technical, critique of the geometry-first approach to special relativity, and defence of the pedagogical value of the alternative. (**)

H. Brown, *Physical Relativity: Space-Time Structure from a Dynamical Viewpoint* (Oxford University Press, 2005), pp. 95–105. A historical and philosophical exploration of relativity, expounding and defending the dynamics-first approach. (**)

J. Earman, *World Enough and Space-Time: Absolute vs Relational Theories of Space and Time* (MIT Press, 1989). The standard graduate-level philosophy text on philosophy of spacetime. (**)

N. Huggett, *Space from Zeno to Einstein: Classic Readings with a Contemporary Commentary* (MIT Press, 1999). Some of the key writings of Newton, Leibniz, and others, accompanied by clear and helpful discussion notes by Huggett.

E. Knox, 'Newtonian Spacetime Structure in Light of the Equivalence Principle', *British Journal for the Philosophy of Science* 65 (2014) pp. 863–80. Technical article on the interpretation of Newtonian gravity. (**)

T. Maudlin, *Philosophy of Physics: Space and Time* (Princeton University Press, 2012). Chapters 1–3 present and defend the geometry-first approach to understanding spacetime and motion. Very clear, from a very specific viewpoint (a viewpoint very different from the one I adopt here). (*)

O. Pooley, 'Substantivalist and Relationist Approaches to Spacetime', in R. Batterman (ed.), *The Oxford Handbook of the Philosophy of Physics* (Oxford University Press, 2013), pp. 522–86. A review article: up-to-date reference for the subject. (**)

E. F. Taylor and J. A. Wheeler, *Spacetime Physics*, 2nd edition (W. H. Freeman, 1992). My favourite of the many introductory books on special relativity; strongly emphasizes the spacetime perspective. (The whole book is available online for free at http://www.eftaylor.com/spacetimephysics/ under a Creative Commons license.) (*)

Philosophy of statistical mechanics (Chapter 4)

D. Albert, *Time and Chance* (Harvard University Press, 1999). One of the most influential philosophy books on statistical mechanics in recent years; idiosyncratic but insightful. (*)

S. Carroll, *From Eternity to Here: The Quest for the Ultimate Theory of Time* (Dutton, 2010). Ambitious but very accessible discussion of time and irreversibility, covering the topics of this chapter but going beyond to more speculative ideas in cutting-edge physics.

R. Feynman, 'The Distinction of Past and Future', *The Character of Physical Law* (MIT Press, 1965), chapter 5. Introduction to the issues from a distinguished physicist.

H. Price, *Time's Arrow and Archimedes' Point* (Oxford University Press, 1996). Extended defence of the idea that our distinction between past and future misleads and confuses us.

L. Sklar, *Physics and Chance: Philosophical Issues in the Foundations of Statistical Mechanics* (Cambridge University Press, 1993). Advanced graduate-level discussion of philosophy of statistical mechanics: a good reference, if slightly out of date by now. (**)

D. Wallace, 'Inferential vs. Dynamical Conceptions of Physics', in O. Lombardi (ed.), *What is Quantum Information?* (Cambridge University Press, 2017). Technical presentation of the inferential/dynamical dichotomy, in both statistical and quantum mechanics. (**)

Philosophy of quantum mechanics (Chapters 5–6)

S. Aaronson, *Quantum Computing Since Democritus* (Cambridge University Press, 2013). Often-insightful, often-infuriating, always-worthwhile, idiosyncratic look at quantum mechanics.

D. Albert, *Quantum Mechanics and Experience* (Harvard University Press, 1994). Reasonably non-technical introduction to the quantum measurement problem, aimed at philosophers.

S. Carroll, *Something Deeply Hidden*: *Quantum Worlds and the Emergence of Spacetime* (Dutton, 2019). Non-technical exposition and defence of the Everett interpretation.

C. Fuchs and A. Peres, 'Quantum Theory Needs No "Interpretation"', *Physics Today* 53 (2000) pp. 70–1. Brief, clear, forceful advocacy of the instrumentalist approach. See also the letters to the editor, and Fuchs and Peres' reply, also in *Physics Today* 53. (*)

C. Fuchs, N. Mermin, and R. Schack, 'An Introduction to QBism with an Application to the Locality of Quantum Mechanics', *American Journal of Physics* 82 (2014) pp. 749–54. (*)

R. Healey, *The Quantum Revolution in Philosophy* (Oxford University Press, 2017). Explores quantum mechanics from a 'pragmatist' point of view fairly closely related to the instrumentalist position. (*)

T. Maudlin, *Philosophy of Physics: Quantum Theory* (Princeton University Press, 2019). Very readable presentation of the measurement problem, followed by detailed (and opinionated) exposition of dynamical-collapse theories, of the de Broglie–Bohm theory, and of the Everett interpretation. Largely ignores probabilistic and/or instrumentalist approaches. (*)

A. Rae, *Quantum Physics: Illusion or Reality?*, 2nd edition (Cambridge University Press, 2004). A general, accessible introduction to the quantum measurement problem and the range of solutions that have been proposed.

R. Penrose, *Shadows of the Mind* (Oxford University Press, 1994), chapters 5–6. Self-contained but reasonably demanding introduction to conceptual problems in QM. (*)

D. Wallace, 'Philosophy of Quantum Mechanics', in D. Rickles (ed.), *The Ashgate Companion to Contemporary Philosophy of Physics* (Ashgate, 2008). Review article, at a reasonably high level (presumes knowledge of quantum mechanics at advanced undergraduate level). (**)

D. Wallace, *The Emergent Multiverse: Quantum Theory According to the Everett Interpretation* (Oxford University Press, 2012). My own book-length treatment of the Everett interpretation; advanced in places. (**)

Index

For the benefit of digital users, indexed terms that span two pages
(e.g., 52–53) may, on occasion, appear on only one of those pages.

Index

RELATIVITY
A Very Short Introduction
Russell Stannard

100 years ago, Einstein's theory of relativity shattered the world of physics. Our comforting Newtonian ideas of space and time were replaced by bizarre and counterintuitive conclusions: if you move at high speed, time slows down, space squashes up and you get heavier; travel fast enough and you could weigh as much as a jumbo jet, be squashed thinner than a CD without feeling a thing - and live for ever. And that was just the Special Theory. With the General Theory came even stranger ideas of curved space-time, and changed our understanding of gravity and the cosmos. This authoritative and entertaining *Very Short Introduction* makes the theory of relativity accessible and understandable. Using very little mathematics, Russell Stannard explains the important concepts of relativity, from E=mc2 to black holes, and explores the theory's impact on science and on our understanding of the universe.

www.oup.com/vsi

RISK
A Very Short Introduction
Baruch Fischhoff & John Kadvany

Risks are everywhere. They come from many sources, including crime, diseases, accidents, terror, climate change, finance, and intimacy. They arise from our own acts and they are imposed on us. In this *Very Short Introduction* Fischhoff and Kadvany draw on both the sciences and humanities to show what all risks have in common. Do we care about losing money, health, reputation, or peace of mind? How much do we care about things happening now or in the future? To ourselves or to others? All risks require thinking hard about what matters to us before we make decisions about them based on past experience, scientific knowledge, and future uncertainties.

www.oup.com/vsi

SCIENTIFIC REVOLUTION
A Very Short Introduction
Lawrence M. Principe

In this *Very Short Introduction* Lawrence M. Principe explores the exciting developments in the sciences of the stars (astronomy, astrology, and cosmology), the sciences of earth (geography, geology, hydraulics, pneumatics), the sciences of matter and motion (alchemy, chemistry, kinematics, physics), the sciences of life (medicine, anatomy, biology, zoology), and much more. The story is told from the perspective of the historical characters themselves, emphasizing their background, context, reasoning, and motivations, and dispelling well-worn myths about the history of science.

www.oup.com/vsi

THE LAWS OF THERMODYNAMICS
A Very Short Introduction
Peter Atkins

From the sudden expansion of a cloud of gas or the cooling of a hot metal, to the unfolding of a thought in our minds and even the course of life itself, everything is governed by the four Laws of Thermodynamics. These laws specify the nature of 'energy' and 'temperature', and are soon revealed to reach out and define the arrow of time itself: why things change and why death must come. In this *Very Short Introduction* Peter Atkins explains the basis and deeper implications of each law, highlighting their relevance in everyday examples. Using the minimum of mathematics, he introduces concepts such as entropy, free energy, and to the brink and beyond of the absolute zero temperature. These are not merely abstract ideas: they govern our lives.

'It takes not only a great writer but a great scientist with a lifetime's experience to explains such a notoriously tricky area with absolute economy and precision, not to mention humour.'

Books of the Year, Observer.

www.oup.com/vsi